Benjamin Bröker

Electronic and structural properties of metal-organic interfaces

Benjamin Bröker

Electronic and structural properties of metal-organic interfaces

Electron donor and acceptor molecules

Südwestdeutscher Verlag für Hochschulschriften

Impressum/Imprint (nur für Deutschland/only for Germany)
Bibliografische Information der Deutschen Nationalbibliothek: Die Deutsche Nationalbibliothek verzeichnet diese Publikation in der Deutschen Nationalbibliografie; detaillierte bibliografische Daten sind im Internet über http://dnb.d-nb.de abrufbar.

Alle in diesem Buch genannten Marken und Produktnamen unterliegen warenzeichen-, marken- oder patentrechtlichem Schutz bzw. sind Warenzeichen oder eingetragene Warenzeichen der jeweiligen Inhaber. Die Wiedergabe von Marken, Produktnamen, Gebrauchsnamen, Handelsnamen, Warenbezeichnungen u.s.w. in diesem Werk berechtigt auch ohne besondere Kennzeichnung nicht zu der Annahme, dass solche Namen im Sinne der Warenzeichen- und Markenschutzgesetzgebung als frei zu betrachten wären und daher von jedermann benutzt werden dürften.

Coverbild: www.ingimage.com

Verlag: Südwestdeutscher Verlag für Hochschulschriften GmbH & Co. KG
Dudweiler Landstr. 99, 66123 Saarbrücken, Deutschland
Telefon +49 681 37 20 271-1, Telefax +49 681 37 20 271-0
Email: info@svh-verlag.de

Approved by: Berlin, Humboldt Universität zu Berlin, Dissertation, 2010

Herstellung in Deutschland:
Schaltungsdienst Lange o.H.G., Berlin
Books on Demand GmbH, Norderstedt
Reha GmbH, Saarbrücken
Amazon Distribution GmbH, Leipzig
ISBN: 978-3-8381-2946-4

Imprint (only for USA, GB)
Bibliographic information published by the Deutsche Nationalbibliothek: The Deutsche Nationalbibliothek lists this publication in the Deutsche Nationalbibliografie; detailed bibliographic data are available in the Internet at http://dnb.d-nb.de.

Any brand names and product names mentioned in this book are subject to trademark, brand or patent protection and are trademarks or registered trademarks of their respective holders. The use of brand names, product names, common names, trade names, product descriptions etc. even without a particular marking in this works is in no way to be construed to mean that such names may be regarded as unrestricted in respect of trademark and brand protection legislation and could thus be used by anyone.

Cover image: www.ingimage.com

Publisher: Südwestdeutscher Verlag für Hochschulschriften GmbH & Co. KG
Dudweiler Landstr. 99, 66123 Saarbrücken, Germany
Phone +49 681 37 20 271-1, Fax +49 681 37 20 271-0
Email: info@svh-verlag.de

Printed in the U.S.A.
Printed in the U.K. by (see last page)
ISBN: 978-3-8381-2946-4

Copyright © 2011 by the author and Südwestdeutscher Verlag für Hochschulschriften GmbH & Co. KG and licensors
All rights reserved. Saarbrücken 2011

Electronic and structural properties of interfaces between electron donor & acceptor molecules and conductive electrodes

DISSERTATION

zur Erlangung des akademischen Grades

Dr. rer. nat.

im Fach Physik

eingereicht an der
Mathematisch-Naturwissenschaftlichen Fakultät I
Humboldt-Universität zu Berlin

von
Herrn M.Sc. Benjamin Bröker

Präsident der Humboldt-Universität zu Berlin:
Prof. Dr. Christoph Markschies

Dekan der Mathematisch-Naturwissenschaftlichen Fakultät I:
Prof. Dr. Andreas Herrmann

Gutachter:
1. Prof. Dr. Norbert Koch
2. Prof. Dr. Jean-Jacques Pireaux
3. Prof. Dr. Antoine Kahn

Tag der mündlichen Prüfung: 25.10.2010

Abstract

The present work is embedded in the field of organic electronics, where charge injection into any kind of device is critically determined by the electronic and structural properties of the interfaces between the electrodes and the conjugated organic materials (COMs). Three main topics are addressed: energy level tuning with new or so far unexplored strong electron (i) donor and (ii) acceptor materials and (iii) the density dependent re-orientation of a molecular monolayer and its impact on the energy level alignment. To study these topics a complementary-technique approach was necessary, including ultra-violet photoelectron spectroscopy (UPS), X-ray photoelectron spectroscopy (XPS) and reflection absorption infrared spectroscopy (RAIRS). Moreover, valuable additional information was obtained from density functional theory (DFT) modeling, which was available through collaboration.

(i) A concept of optimizing the energy level alignment at interfaces with strong molecular acceptors was extended to donor materials and thus successfully transfered from the anode to the cathode side of the device. Also in this case, charge transfer between the donor molecule and the metal electrode leads to a chemisorbed molecular monolayer. Due to the dipole across the interface, the work function of the electrode is reduced by up to 2.2 eV. Consequently, a reduced electron injection barrier (Δ_e) into subsequently deposited electron transport materials is achieved (up to 0.8 eV). (ii) A yet unexplored strong electron acceptor material [i.e. hexaaza-triphenylene-hexacarbonitrile (HATCN)] is thoroughly investigated on model electrode and application relevant indium tin oxide (ITO) surfaces. On both HATCN shows superior performance as electron acceptor material compared to presently used materials for energy level tuning (e.g. work function modification and hole injection barrier reduction (Δ_h) by up to 1 eV). (iii) Also with HATCN, the orientation of a molecular monolayer is observed to change from a face-on to an edge-on conformation depending on layer density, facilitated through specific interactions of the peripheral molecular cyano groups with the metal. This is accompanied by a re-hybridization of molecular and metal electronic states, which significantly modifies the interface and surface electronic properties. All findings presented are valuable for the understanding of electrode-COM interfaces and will thus help in advancing the field of organic electronics.

Zusammenfassung

Die vorliegende Arbeit behandelt Fragestellungen aus der Organischen Elektronik, in der die Ladungsträgerinjektion in alle Arten von Bauteilen kritisch von der elektronischen und morphologischen Struktur der Grenzflächen zwischen Elektrode und den konjugierten organischen Molekülen (KOM) abhängt. Drei Themen wurden näher betrachtet: die Energieniveauanpassung mit neuen und bis heute unerforschten starken (i) Elektronendonatoren und (ii) -akzeptoren und (iii) die dichteabhängige Umorientierung einer molekularen Monolage und ihr Einfluss auf die elektronische Struktur. Um diese Themen zu analysieren war ein breiter experimenteller Ansatz nötig, bei dem ultraviolette Photoelektronenspektroskopie, Röntgenphotoelektronenspektroskopie und Reflektionsabsorptionsinfrarotspektroskopie angewandt wurden. Weitere wertvolle Informationen konnten durch Modellierung mit Dichtefunktionaltheory gewonnen werden, die über Kollaborationen zur Verfügung standen.

(i) Das Konzept der optimierten Energieniveauanpassung mit starken Elektronenakzeptoren konnte auf Donatoren erweitert und damit erfolgreich von der Anode zur Kathode transferiert werden. Auch hier führte der Ladungstransfer vom Molekül zur Metallelektrode zu einem Dipol über die Grenzfläche, womit die Austrittsarbeit um bis zu 2.2 eV reduziert werden konnte. Als Resultat konnte die Elektroneninjektionsbarriere in nachfolgende Elektronentransportmaterialien entscheidend verringert werden (bis zu 0.8 eV). (ii) Ein bis dato unerforschter starker Elektronenakzeptor [hexaaza-triphenylene-hexacarbonitrile (HATCN)] wurde vollständig auf Modell- und anwendungsrelevanten Elektroden charakterisiert. In beiden Fällen zeigte HATCN eine bessere Performance verglichen mit derzeit üblichen Materialien zur Energieniveauanpassung (starke Austrittsarbeitsanhebung und Verringerung der Lochinjektionsbarriere um bis zu 1.0 eV). (iii) Zusätzlich konnte mit HATCN gezeigt werden, dass eine liegende molekulare Monolage durch Erhöhung der Moleküldichte auf der Oberfläche in eine stehende Monolage umgewandelt werden kann. Dies führte zu einer änderung der chemischen Bindung zum Metall und damit zu einer starken Modifikation der elektronischen Struktur der Grenzfläche. Die Ergebnisse der vorliegenden Arbeit liefern wertvolle Informationen für das Verständnis der Grenzfläche zwischen Elektrode und KOM und tragen damit zum Fortschritt der Organischen Elektronik bei.

Contents

Abbreviations — ix

1. Introduction — 1

2. Fundamentals of metal/organic interfaces — 5
  2.1. Electronic structure of metal/organic interfaces — 6
    2.1.1. Conductive surfaces — 6
    2.1.2. Isolated conjugated organic molecules — 8
    2.1.3. Molecular organic solids — 8
    2.1.4. Energy level alignment at interfaces between conductive surfaces and conjugated organic molecules — 10
    2.1.5. Charge injection — 16
  2.2. Morphological structure of metal/organic interfaces — 19
    2.2.1. Growth — 19

3. Experimental methods — 23
  3.1. Photoelectron spectroscopy — 23
    3.1.1. Theoretical background — 27
    3.1.2. Line width and shape — 31
    3.1.3. Selection rules — 31
    3.1.4. Many-body effects — 32
    3.1.5. Auger electron spectroscopy — 35
    3.1.6. Quantitative analysis of core level spectra — 35
  3.2. Infrared spectroscopy — 36
    3.2.1. Molecular vibrations — 36
    3.2.2. Vibrations at surfaces — 40
    3.2.3. Information content of vibrational modes of adsorbed molecules — 46
    3.2.4. Fourier-transform infrared spectroscopy — 48

4. Materials and experimental setups — 51
  4.1. Materials — 51
    4.1.1. Molecular electron donor and acceptor materials — 51
    4.1.2. Hole and electron transport materials — 53
  4.2. Substrates — 54
    4.2.1. Metal single crystals — 54
    4.2.2. Indium tin oxide — 56
  4.3. Experimental setups — 56
    4.3.1. Photoelectron spectroscopy experiments at BESSY II — 56
    4.3.2. Photoelectron spectroscopy experiments at HASYLAB and Humboldt-University — 59

Contents

    4.3.3. Reflection absorption infrared spectroscopy experiments at Zernike Institute for Advanced Materials 60
4.4. Experimental details 62
    4.4.1. Sample preparation 62
    4.4.2. UPS and XPS experiments 62
    4.4.3. RAIRS experiments 63
4.5. Theoretical calculations 63

5. Results and Discussion   67

5.1. Initial screening of donor & acceptor molecules 68
5.2. MV0 on coinage metals 70
    5.2.1. Introduction 70
    5.2.2. Valence electronic structure 72
    5.2.3. Core level analysis 79
    5.2.4. RAIRS experiments of MV0 on Ag(111) 81
    5.2.5. DFT results 82
    5.2.6. Electron injection barrier tuning with MV0 interlayers 84
    5.2.7. Conclusions 92
5.3. NMA on coinage metals 93
    5.3.1. Introduction 93
    5.3.2. Photoelectron spectroscopy at interfaces to metals 93
    5.3.3. Electron injection barrier tuning with NMA interlayers 102
    5.3.4. Conclusions 104
5.4. The model acceptor F4-TCNQ adsorbed on Ag(111) 105
    5.4.1. Introduction 105
    5.4.2. Valence electronic structure 106
    5.4.3. DFT results 108
    5.4.4. RAIRS results 109
    5.4.5. Conclusion 111
5.5. HATCN on Ag(111) 113
    5.5.1. Introduction 113
    5.5.2. Evolution of the work function 114
    5.5.3. TDS and RAIRS experiments 115
    5.5.4. Theoretical modeling of HATCN on Ag(111) 120
    5.5.5. UPS valence band spectra of HATCN on Ag(111) 121
    5.5.6. α−NPD on HATCN 123
    5.5.7. Conclusions 123
5.6. HATCN adsorbed on Cu(111) and Au(111) 124
    5.6.1. Valence electronic structure 125
    5.6.2. RAIRS results 128
    5.6.3. Conclusion 131
5.7. Tuning the hole injection barrier from indium tin oxide into α-NPD 133
    5.7.1. Introduction 133
    5.7.2. F4-TCNQ on ITO 133
    5.7.3. HATCN on ITO 135
    5.7.4. Hole injection barriers from ITO into α−NPD 136
    5.7.5. Conclusion 138

6. Summary and Outlook **141**

A. Appendix **145**
   A.1. Character and Correlation Tables 145
   A.2. Image dipole theory and the surface selection rule 147
   A.3. Beam energies and photon flux 148
   A.4. C1s spectra of a saturated monolayer of MV0 on Ag(111), Cu(111), and Au(111) 149
   A.5. Additional experimental results for NMA 151
        A.5.1. Full core level spectra of NMA on Au(111), Ag(111), and Cu(111) . . 152
   A.6. Vibrational modes of a thick F4-TCNQ film on Ag(111) 154
   A.7. Fitting of the RAIR spectra for thin films of HATCN on Ag(111) 155
   A.8. Theoretical description of the upright standing HATCN layer on Ag(111) . . 156
   A.9. Auger electron spectroscopy 158
   A.10.STM results for the adsorption of HATCN on Ag(111) and Cu(111) 160
   A.11.Valence electronic structure of α−NPD on ITO 161

Abbreviations

A	Richardson Constant
$\vec{A}$	Vector Potential of the incident Light
$A_{p\parallel}$	Amplitude of the Electric field parallel
	to the plane of incidence and parallel to the surface
$A_{p\parallel}$	Amplitude of the Electric field parallel
	to the plane of incidence and perpendicular to the surface
A_s	Amplitude of the Electric field perpendicular
	to the plane of incidence
$A_{i,p}$	Amplitude of the incoming Electric Field
	polarized parallel to the surface
$A_{r,p}$	Amplitude of the reflected Electric Field
	polarized parallel to the surface
A_x	Integrated Peak area of a specific Core Level Emission
ASF	Atomic Sensitivity Factor
BE	Binding Energy
c	Velocity of Light
CLR	Core Level Region
COM	Conjugated Organic Material
CT	Charge Transfer
C_x	Percentage of a specific Element in a homogeneous Sample
d_{eff}	Effective atomic/molecular Diameter
DFT	Density Functional Theory
DOS	Density Of States
e	Elementary Charge
E	Energy
EA	Electron Affinity
EA_{gas}	Gas Phase Electron Affinity
EDC	Electron Distribution Curve
E_B	Binding Energy
$E_f(\vec{k})$	Energy of an Final State
E_F	Fermi-level

Contents

$E_{G,gas}$	Energy Gap of a Molecule in the Gas Phase
$E_{G,opt}$	Optical Gap of a Molecule in the Solid State
$E_{G,trans}$	Transport Gap of a Molecule in the Solid State
$E_i(\vec{k})$	Energy of an Initial State
$\vec{E}$	Electric Field at the Surface
$\vec{E}_p$	Electric Field at the Surface polarized parallel to the plane of incidence
$\vec{E}_s$	Electric Field at the Surface polarized perpendicular to the plane of incidence
$\vec{E}_i$	Electric Field of the incident Light
$\vec{E}_{i,p}$	Component of $\vec{E}_i$ polarized parallel to the plane of incidence
$\vec{E}_{i,s}$	Component of $\vec{E}_i$ polarized perpendicular to the plane of incidence
$\vec{E}_{i,p\parallel}$	Component of $\vec{E}_{i,p}$ parallel to the surface
$\vec{E}_{i,p\perp}$	Component of $\vec{E}_{i,p}$ perpendicular to the surface
E_{kin}	Electron Kinetic Energy
E'_{kin}	Measured Electron Kinetic Energy
E_{kin,E_F}	Electron Kinetic Energy of the Fermi-level
$E_{kin,HOMO\,onset}$	Electron Kinetic Energy of the HOMO Onset
$E_{kin,SECO}$	Electron Kinetic Energy of the SECO
$E_{(P+)}$	Positive Polarization Energy of a Molecule in the Solid State
$E_{(P-)}$	Negative Polarization Energy of a Molecule in the Solid State
$E(v)$	Energy of a Vibration with Quantum Number v
E_{VAC}	Vacuum-level
$E_{VAC,\infty}$	Vacuum-level at Infinity
EMT	Electron Transport Material
F	external applied Electric Field
F4-TCNQ	2,3,5,6-tetrafluoro-7,7,8,8-tetracyano-quinodimethane
FWHM	Full Width at Half Maximum
H	Hamilton Operator
H'	Perturbation Operator
$h\nu$	Photon Energy
HATCN	hexaazatriphenylene-hexacarbonitrile
HOMO	Highest Occupied Molecular Orbital
HTM	Hole Transport Material
I	Intensity
I^{ext}	External Emission Current
$I^{int}(E'_{kin}, h\nu, \vec{k})$	Internal Electron Current

ID	Interface Dipole
IE	Ionization Energy
IE_{gas}	Gas Phase Ionization Energy
IR	Infrared
ITO	Indium Tin Oxide
j	Current Density
k	imaginary Part of the Complex Refractive Index
k_B	Boltzmann Constant
k_s	Spring Constant
$\vec{k}$	Wave Vector
LEED	Low Energy Electron Diffraction
LUMO	Lowest Unoccupied Molecular Orbital
m_e	Electron Mass
m_{red}	Reduced Mass
$M_{i \to f}$	Transition Dipole Moment from State i to f
MCT	Mercury Cadmium Telluride
MV+1	singly oxydized Methyl Viologen
MV0	neutral Methyl Viologen
n	real Part of the Complex Refractive Index
N_0	Number of intrinsic Charges
n_x	Density of a specific Element in a certain Volume
$n(z)$	Positive Uniform Background Charge
NMA	9,9'-ethane-1,2-diylidene-bis(N-methyl-9,10-dihydroacridine)
OFET	Organic Field Effect Transistor
OLED	Organic Light Emitting Diode
OPVC	Organic Photovoltaic Cell
$\underline{P}$	Momentum Operator
$P_{v \to v\prime}$	Transition Probability from State with v to $v\prime$
P1	Pinning Level above the Fermi-level
P2	Pinning Level below the Fermi-level
PES	Photoelectron Spectroscopy
QCM	Quartz Crystal Microbalance
r_0	Equilibrium Distance
R_0	Reflectivity of the uncovered Surface
R_A	Reflectivity of the Adsorbate covered Surface
RAIRS	Reflection Absorption Infrared Spectroscopy
r_p	Reflection coefficient of the electric Field

Contents

	polarized parallel to the plane of incidence
R_p	Reflected Intensity of the electric Field
	polarized parallel to the plane of incidence
r_s	Reflection coefficient of the electric Field
	polarized perpendicular to the plane of incidence
R_s	Reflected Intensity of the electric Field
	polarized perpendicular to the plane of incidence
SAM	Self Assembled Monolayer
SCLC	Space-Charge-Limited-Current
SD	Surface Dipole
SECO	Secondary Electron Cut-Off
STM	Scanning Tunneling Microscope/Microscopy
T	Temperature
$T(E'_{kin}, \vec{k})$	Electron Transport Function
TDS	Thermal Desorption Spectroscopy
TTC	Tetratetra-contane
UHV	Ultra High Vacuum
U_{ext}	External applied Bias
U_{SECO}	Applied Bias during SECO measurement
UPS	Ultraviolet Photoelectron Spectroscopy
v	Vibrational Quantum Number
VASP	Vienna Ab-initio Simulation Package
VB	Valence Band
VR	Valence Region
$X(E'_{kin}, \vec{k})$	Electron Escape Function
XPS	X-ray Photoelectron Spectroscopy
W	Transition Probability per Unit Time
α	Angle between Escaping Electron and Surface Normal
β	"ideality" factor of the Schottky-Diode
γ	Angle between incident/reflected Light and the Surface Normal
γ_{eh}	Ratio of injected Electrons and Holes
δ	Optical Path Difference
δ_e	Injection Barrier reduction
δ_p	Phase Shift of the electric Field polarized parallel to the plane of incidence
δ_s	Phase Shift of the electric Field polarized perpendicular to the plane of incidence
Δ_h	Hole Injection Barrier
Δ_e	Electron Injection Barrier

$\Delta\Phi$	Work Function Change
ΔR	Reflectivity Change
ϵ	Complex dielectric Constant
η_{ex}	Proportion of Excitons that can Decay radiatively
η_{oc}	Efficiency of Light out-coupling
η_{pl}	Efficiency of radiative Decay
η_Q	Quantum Efficiency
θ	nominal Film Thickness
θ_{tr}	Parameter accounting for the Density and Depth of Traps
λ	Mean-Free Electron Path
μ	Chemical Potential
$\underline{\mu}$	Dipole Moment Operator
μ_b	Bulk Chemical Potential
μ^{COM}	Mobility in a COM
ν	Frequency
$\tilde{\nu}$	Wavenumber
$\rho(z)$	negative Charge Density spilling out into Vacuum
φ	Phase
Φ	Work Function
Φ_A	Analyzer Work Function
φ_{inner}	inner Electrostatic Potential
φ_{outer}	outer Electrostatic Potential
φ_{field}	Potential of the applied Electric Field
φ_{image}	Image Potential
Φ_{light}	Scalar Potential of the incident Light
Φ_{mod}	modified Work Function
Φ_S	Surface/Substrate Work Function
$\varphi(z)$	electrostatic Potential in z-Direction
Ψ_i	Electron Initial State
Ψ_f	Electron Final State
$\Psi(v)$	Vibrational Initial State
$\Psi(v\prime)$	Vibrational Final State

Table 1.: Acronym, chemical formula, name, and molecular weight of the different donor and acceptor materials initially screened in this work.

Acronym	Chemical formula	Name	Molecular weight (g mol^{-1})
MV0	$C_{12}H_{14}N_2$	1,1'-dimethyl-4,4'-bipyridinylidene	186
BBTHC	$C_{30}H_{25}N_1$	bisbenzothienylhexyl-carbazole	463
NMA	$C_{30}H_{24}N_2$	9,9'-ethane-1,2-diylidene-bis(N-methyl-9,10-dihydroacridine)	412
BEDT-TTF	$C_{10}H_8S_8$	bis(ethylene-dithiolo)-tetrathiafulvalene	384
BEDO-TTF	$C_{10}H_8O_4S_4$	bis(ethylene-dioxy)-tetrathiafulvalene	320
HPyB	$C_{30}H_{24}N_6$	hexapyrrolyl-benzene	468
Ome-HPB	$C_{30}H_{24}N_6$	octadecamethoxy-hexaphenylbenzene	1074
HMT	$C_{20}H_{22}O_6$	hexamethoxy-tolan	358
PD	$C_{22}H_{12}O_2$	pentacene-dion	308
PT	$C_{22}H_{10}O_4$	pentacene-tetraone	338
PyT	$C_{16}H_6O_4$	pyrene-tetraone	262
COHON	$C_{24}H_6O_6$	coronene-hexaone	390
TNFCN	$C_{16}H_5N_5O_6$	dicyanomethyl-trinitrofluorenone	263
Br-PyT	$C_{16}H_4Br_2O_4$	dibromo-pyrene-tetraone	420
NO2-PyT	$C_{16}H_4N_2O_8$	dinitro-pyrene-tetraone	352
F4-TCNQ	$C_{12}F_4N_4$	2,3,5,6-tetrafluoro-7,7,8,8-tetracyano-quinodimethane	276
HATCN	$C_{18}N_{12}$	1,4,5,8,9,12-hexaazatriphenylene-hexacarbonitrile	384

Chapter 1.

Introduction

The field of organic electronics, which includes several classes of devices like organic light emitting diodes (OLEDs), photovolthaic cells (OPVCs), and field effect transistors (OFETs), dates back to early experiments on the ground and excited states of model organic semiconductors in the 1960s [1]. Further progress was made by studying the properties of conductive conjugated polymers and ultrapure polyacene crystals [2, 3]. Additional momentum was gained after the demonstration of electroluminescence of organic materials by Tang and VanSlyke more than two decades ago [4]. Since then research in university and industry has strongly increased. The interest is driven by the prospects of low weight, flexible, and low cost electronics, which mark the main advantages over traditional silicon based electronics [5]. Moreover, organic molecules present a system, whose electronic, optical, and structural properties can be fine-tuned over a wide range by chemical synthesis, which is one of *the* key advantages [6]. Often, a strong interdependence between these properties is found, which is not always beneficial for the actual device performance. Even though wide research is carried out in the field of organic electronics, these general relationships between structure and electronic properties are only scarcely understood. Furthermore, the interfaces between the electrodes and the conjugated organic material (COM) in organic electronic devices can still be identified as one of the key areas for device improvement [7, 8]. Consequently, considerable efforts are being aimed at understanding the morphological and electronic structure right at the interface to facilitate charge carrier injection/extraction and increase the quantum efficiency of the devices. To improve this issue, several strategies towards adjusting the electronic levels of the electrode-COM interface exist, one of them being the adsorption of molecular acceptors and donors that undergo a charge-transfer-type reaction with the electrode materials. Due to this reaction, the work function (Φ) of the electrode is modified as linear function of donor/acceptor coverage (θ) as shown in Fig. 1.1. Note that in this illustrative picture dipole-dipole interactions are neglected and growth is assumed to be layer-by-layer. The maximum work function modification is thus reached, when the first molecular layer is fully closed. With increasing donor/acceptor coverage, the work function remains constant. The energy levels of subsequently deposited hole-/electron-transport materials (HTM and ETM, respectively) are then

Chapter 1. Introduction

aligned to this modified electrode work function. In the simplest case, this allows continuous tuning of the hole- and electron injection barriers (Δ_h/Δ_e) between the two extreme cases: a bare electrode and the full donor/acceptor monolayer (see Fig. 1.1). This approach has already proven to be successful. By application of the strong electron acceptor F4-TCNQ, the hole injection barriers into subsequently deposited N,N'-diphenyl-N,N'-bis(1-naphthyl)-1,1'-biphenyl-4,4'-diamine (α-NPD), sexithiophene (6T), and sexiphenyl (6P) could be lowered by as much as 1.2 eV [9]. As donor molecules, tetrakis(dimethylamino)ethylene (TDAE) and cobaltocene derivatives were successfully used to lower the work function of metal electrode surfaces [10, 11, 12, 13]. A decrease of Δ_e into subsequently deposited ETM layers was suggested on the basis of current-voltage characteristics measured for a TDAE modified device.

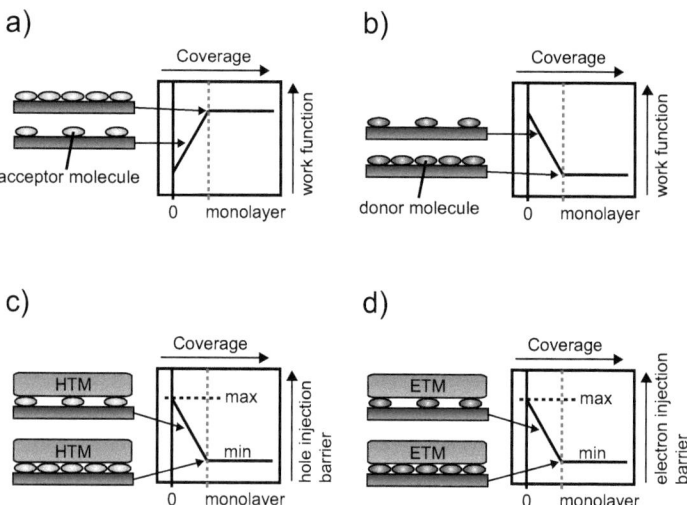

Figure 1.1.: a) and b) schematic plots of the work function as a function of coverage for (sub-) monolayer to multilayer coverages of electron a) acceptor and b) donor molecules (neglecting dipole-dipole interactions and island growth). c) and d) schematic plot of the injection barrier from the substrate into a thick layer of a c) HTM and a d) ETM as a function of c) acceptor and d) donor pre-coverage.

However, the donor/acceptor molecules used so far for decreasing Δ_h and Δ_e have virtually no application potential, as their molecular weight is too low for integration in an industrial production process (i.e., the temperature for vacuum evaporation is below 100°C). This leads to cumbersome processability and potential short device lifetimes because of diffusion throughout the active organic material [14, 15]. In order to overcome these limitations, new and strong donor and acceptor materials with higher molecular weight are needed. Their careful analysis is the main objective of the present work, which was done within the framework of the European Communion project "ICONTROL". In order to find suitable molecules, a large variety of different molecular structures was subjected to an initial selection process. This comprised the adsorption of the molecules on coinage metal crystals and their analysis using UPS. As a first design guide structures based on already known electron donor and acceptor molecules were used. The selection criterion for the molecules to be chosen for further analysis was their work function modification potential. Selected molecules were subsequently investigated more in-depth using ultraviolet and X-ray PES, infrared spectroscopy, and to some extend other complementary experimental techniques to access the electronic and morphological structure at the metal/molecule interface. Most of the experiments were accompanied by extensive theoretical modeling using density functional theory (DFT) methodology to achieve additional insights (this was done in the group of Egbert Zojer at the Technical University (TU) of Graz, Austria). To clarify if the work function modification translates into a lowering of the charge injection barrier, additional UPS experiments employing prototypical hole- and electron transport materials were carried out.

This work is composed as follows: In Chapter 2 fundamental aspects of conductive electrode surfaces and conjugated organic molecules will be illustrated. The chapter will start with the properties of the separate entities: conductive electrodes (Sec. 2.1.1), isolated COMs (Sec. 2.1.2), and molecular assemblies (Sec. 2.1.3). The chapter will then continue with the description of interfaces between the electrodes and COMs and several models describing the energetics will be outlined (Sec. 2.1.4). Additional attention will be given to the growth of COMs on electrode surfaces (Sec. 2.2.1) and the charge injection from the electrodes into the organic materials (Sec. 2.1.5). Chapter 3 gives a comprehensive theoretical background of the PES and RAIRS techniques used within this work. The materials that have been investigated are described in chapter 4, where also details of all experimental setups are given. Additionally, the experimental details and methods of data analysis are given. At the end of chapter 4 a brief introduction to the DFT methodology used for the theoretical description will be outlined (Sec. 4.5). Chapter 5 presents the main part of this work. Here, the results of the experimental studies are presented and discussed in detail. The chapter is divided into seven parts. In the first part (Sec. 5.1) the results of the molecular selection process are shown and trends in the work function modification are examined. In the second part (Sec. 5.2), the adsorption of MV0 on metallic substrates is investigated. MV0 represents a strong electron donor, but has only a small molecular weight. Consequently, its application

Chapter 1. Introduction

potential is rather limited, but it serves as a donor benchmark system because of its strong work function modification and large Δ_e reduction at the interfaces to subsequently deposited C_{60} and 8-hydroxyquinoline aluminum (Alq_3). To improve the situation, a larger molecular donor (NMA) is investigated in the following part (Sec. 5.3). Its work function modification potential is not as strong as that of MV0, however, a continuous tuning of Δ_e is achieved. The molecule thus presents a viable starting point for the synthetic route to strong and large donor materials. In literature, the acceptor material F4-TCNQ has been investigated on gold and copper surfaces [9, 16]. In Sec. 5.4 the adsorption of the acceptor is analyzed on silver surfaces to complete the set of coinage metal substrates and have a full benchmark system. As larger acceptor molecule, the adsorption of HATCN is investigated first on silver surfaces (Sec. 5.5). Here, a change in orientation of the first molecular layer from a face-on to an edge-on conformation is observed depending on layer density. The adsorption of HATCN on copper and gold is described in the following section (Sec. 5.6). Since HATCN strongly increases the work function on silver and copper substrates and shows potential to reduce Δ_h into subsequently deposited α−NPD, its adsorption on application relevant ITO is explored in to Sec. 5.7. The results are compared to the adsorption of F4-TCNQ on ITO. Thereby it is found that the Δ_h decrease is larger for HATCN compared to F4-TCNQ pre-covered ITO. Finally, all results are summarized in chapter 6 and an outlook is given based on the experimental and theoretical findings obtained within this work.

Chapter 2.

Fundamentals of metal/organic interfaces

In this chapter the most relevant parameters characterizing metal/organic interfaces will be outlined. First, as an example of an organic electronic device, the OLED will be elucidated and the importance of the interfaces between conductive electrode and organic material will be highlighted. The chapter will continue with a detailed description of the separated entities starting with the properties of conductive surfaces. Next, the properties of isolated molecules will be illustrated and then extended to molecular solids. Finally, molecules in intimate contact with conductive surfaces will be discussed with respect to the energy level alignment, molecular growth, and charge injection across the interface.

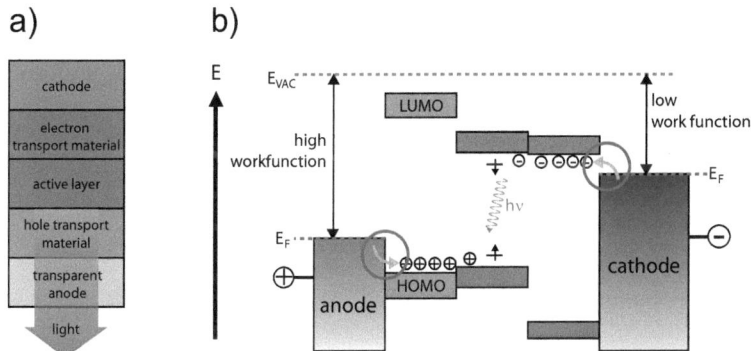

Figure 2.1.: a) layer stack of an organic light emitting diode (OLED). b) frontier energy level diagram of the OLED stack shown in a). The red circles indicated the hole (anode-hole transport material interface) and electron injection barrier (cathode-electron transport material interface) of the device.

The layer stack of an OLED device is shown in Fig. 2.1a. In general, the organic layers are sandwiched between conductive anode and cathode. One of the two electrodes needs to be transparent to be able to extract light from the device. Current state-of-the-art OLEDs

Chapter 2. Fundamentals of metal/organic interfaces

feature transparent conductive oxides like ITO or highly conductive polymers as anode materials. In the presented case, the organic layer is comprised of a HTM, ETM and an active material, in which the light is generated. The corresponding frontier energy level diagram of the stack shown in Fig. 2.1a is plotted in Fig. 2.1b. By applying a voltage between both electrodes, holes are injected from the Fermi level of the anode into the charge transport levels located energetically above the highest occupied molecular orbital (HOMO) of the HTM. Similarly, electrons are injected from the Fermi level of the cathode into charge transport levels located slightly below the lowest unoccupied molecular orbital (LUMO) of the ETM. The charge transport levels correspond to the positive (hole) and negative (electron) polaron levels of the organic materials. In other words, the molecules, which transport the charges, are ionized. This is what makes ionizing investigation techniques such as photoelectron spectroscopies well suited for the analysis of the charge transport levels (see Sec. 2.1.3). In both cases the holes and electrons have to overcome barriers (Δ_h and Δ_e) in order to be injected (see Sec. 2.1.5). After their injection, the charge carriers travel in direction of the opposite electrodes due to the potential drop across the organic layers. In the active material holes and electrons form excitons, which then eventually recombine radiatively. The number of emitted photons per injected electron is termed the quantum efficiency η_Q of the device. It is given by [17]:

$$\eta_Q = \gamma_{eh} \cdot \eta_{ex} \cdot \eta_{pl} \cdot \eta_{oc} \tag{2.1}$$

where γ_{eh} represents the ratio of injected electrons and holes, η_{ex} the proportion of excitons that can decay radiatively, η_{pl} the efficiency of radiative decay, and η_{oc} the efficiency of light out-coupling, which is the number of photons that actually leave the device. The power efficiency η_P of the device, which also takes into account the wavelength-dependent response of the human eye, is proportional to the quantum efficiency. Its optimization is consequently crucial for the OLED device performance. The concept of injection barrier lowering introduced in Sec. 1 tackles the optimization via γ_{eh}. The maximum value that can be taken by γ_{eh} is 1, which means one injected electron per hole. Besides being influenced by the injection barriers γ_{eh} is also affected by the mobility of the charge carriers in the organic materials.

2.1. Electronic structure of metal/organic interfaces

2.1.1. Conductive surfaces

Conductive surfaces represent the outermost part of the electrodes used in organic electronic devices as shown in the previous section. Filled and empty states are separated by the Fermi level E_F and the difference between vacuum-level E_{VAC} and E_F is the work function Φ. Φ depends on the chemical potential μ of the electrons in the bulk and the change in electrostatic

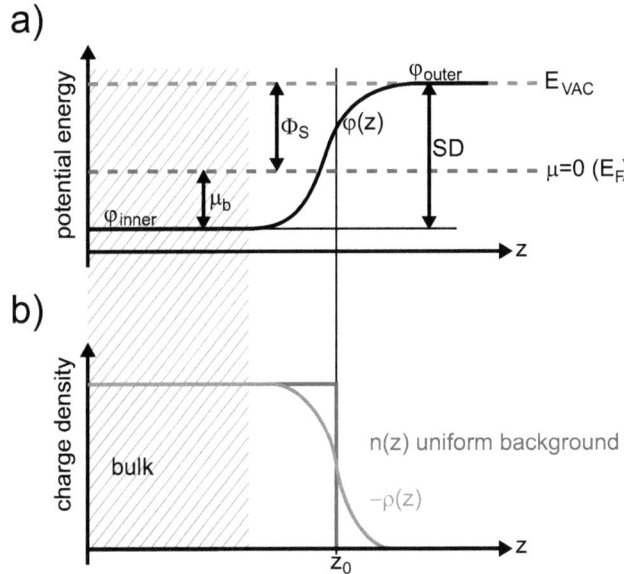

Figure 2.2.: a) Potential energy $\varphi(z)$ as a function of distance z across a metal/vacuum interface. b) positive uniform background charge $n(z)$ and negative electron density $\rho(z)$ as a function of vertical distance z relative to a metal surface.

potential $\varphi(z)$, when going from the bulk across the interfaces into vacuum, as shown in Fig. 2.2. Due to the non-vanishing probability of electrons residing outside the metal surface (tunneling), a finite negative charge density $\rho(z)$ spills out into vacuum; as calculated by Lang and Kohn [18], this spill-out can extend to several Å. As a consequence of overall charge neutrality, negative charge density is missing inside the metal (see Fig. 2.2b). The material's positive charge density created by the atomic nuclei $n(z)$ can be described in the jellium model as a positive uniform background charge, which shows a step-like behavior as it drops from a constant value inside the metal to zero at the interface [18]. The difference of inner potential energy φ_{inner} and the potential energy outside of the metal φ_{outer} defines the absolute change of the electron potential energy. The corresponding potential change is often termed the surface dipole SD. φ_{inner} of the electrons is usually set to zero in the jellium model [19]. Then the difference between φ_{inner} and the Fermi level equals the bulk chemical potential μ_b of the electrons [19]. The surface work function of a metal is according

Chapter 2. Fundamentals of metal/organic interfaces

to Fig. 2.2a:
$$\Phi_S = -\mu_b + SD \tag{2.2}$$

Because μ_b is a bulk property, it is not changed by adsorbates, however, the SD, as determined by the spill-out of electrons into vacuum, can easily be modified by adsorbed species, and thus Φ_S as well. This relationship is nicely illustrated for the adsorption of Xe on various metal substrates in Ref. [20]. This phenomenon is often called electron "push back" effect, because the surface electron density spilling out into vacuum [$\rho(z)$ above the surface (c.f. Fig. 2.2)] is pushed back into the metal by Coulomb (Pauli) repulsion of the adsorbate's electron density. Further details will be discussed in Sec. 2.1.4.

2.1.2. Isolated conjugated organic molecules

Conjugated organic molecules are usually closed shell systems in which $\pi-$ and $\sigma-$bonds predominate. The molecular orbitals are occupied by electrons up to the HOMO, which is separated by an energy gap (in the gas phase) $E_{G,gas}$ from the LUMO. The schematic energy level diagram in Fig. 2.3a shows the vertical ionization energy IE_{gas} and electron affinity EA_{gas} of a molecule in the gas phase as derived by ionizing analysis techniques such as PES or inverse photoemission. IE_{gas} is the energy needed to remove an electron from the neutral molecules HOMO and bring it to the vacuum-level at infinite distance $E_{VAC,\infty}$, and EA_{gas} is the energy gained upon adding an electron to the neutral molecules LUMO from $E_{VAC,\infty}$. In both cases, the molecules are ionized. Since a common vacuum-level E_{VAC} is already reached close above the surfaces (a few Å) and the freedom of choosing the origin of the potential energy, $E_{VAC,\infty} = E_{VAC}$ are generally set to zero. Thus both, IE_{gas} and EA_{gas}, are energy differences between an initial and a final state and well defined for a single isolated molecule.

2.1.3. Molecular organic solids

Depending on the extent of periodic long range order of molecules in a solid, one can distinguish molecular single crystals, polycrystalline, and amorphous solids [21]. Usually the inter-molecular forces (mainly van-der-Waals forces) in molecular crystals are weak, due to the closed shell nature of the molecules and the small spatial overlap of the electron wave functions of neighboring molecules. By and large, the electronic properties of individual molecules are preserved in the solid. The introduction of a charge in the solid leads to significant electronic and structural relaxations, which are important to consider for charge transport. Consequently, charges are transported through the molecular crystal not as free particles, but "dressed" in a polarization cloud [22], the corresponding energy levels are called polaron levels. The polaron levels of the ionized molecule in the solid (as shown in Fig. 2.3b) differ

2.1. Electronic structure of metal/organic interfaces

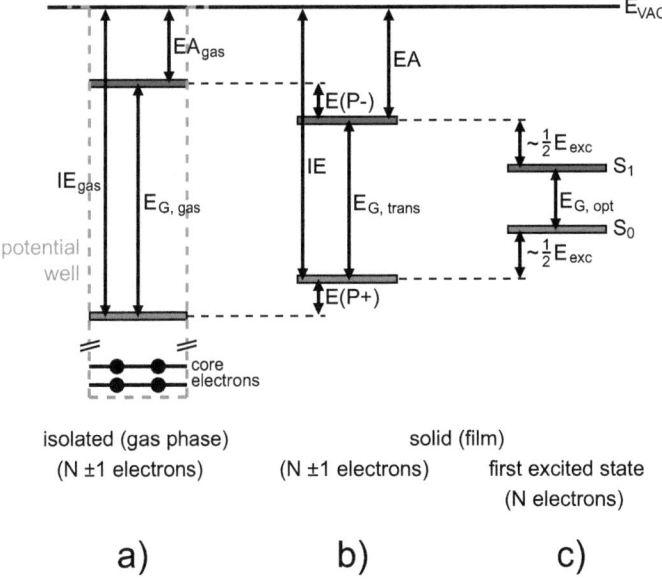

Figure 2.3.: Schematic frontier energy levels of a) an isolated ionized molecule in the gas phase, with IE_{gas}, EA_{gas} and the energy gap $E_{G,\,gas}$. Moreover, the potential well for the molecule is shown including core electrons. b) Polaron levels of an ionized molecule in a solid (condensed film), where the polarization of the surrounding media decreases the IE_{gas} by the positive polarization energy $E_{(P+)}$, while EA_{gas} is increased by the negative polarization energy $E_{(P-)}$. The energetic difference between both polaron levels is the transport gap $E_{G,trans}$. In c) the fundamental optical transition is shown, creating a Frenkel-exciton. This transition defines the optical gap $E_{G,opt}$ and the difference between $E_{G,opt}$ and $E_{G,trans}$ is the exciton binding energy E_{exc}.

from the levels of the isolated ionized molecule (Fig. 2.3a) by the positive $E_{(P+)}$ and negative polarization energy $E_{(P-)}$ caused by the polarizability of the surrounding molecules in the solid. They can be on the order of several hundred meV [23, 24]. The energy difference between the two polaron levels in the solid is the transport gap $E_{G,trans}$. The first fundamental optical transition is depicted in Fig. 2.3 as well, which creates a Frenkel-exciton. In this case the hole and electron are located on one molecule. The minimal energy required for this transition equals the optical gap $E_{G,opt}$. As can be deduced from Fig. 2.3b and c, the difference between $E_{G,trans}$ and $E_{g,opt}$ gives the charge separation energy or exciton binding

Chapter 2. Fundamentals of metal/organic interfaces

energy (E_{exc}), which is substantial in organic semiconductor materials and has been reported to be as large as 1.5 eV [24, 25, 26]. This is one of the major differences between organic and inorganic semiconductor materials, where in the latter the charge separation energy is generally below $k_B T$ [27].

Different crystal faces of a metal single crystal have different work function values [28]. The prerequisite for this observation is that the surface is laterally extended, because the spatial region above the sample where E_{VAC} is raised due to the SD reaches farther away from the surface with increasing sample size, i.e., the sample geometry defines the electrostatic potential landscape [29]. Similarly, IE and EA of ordered molecular solids have been shown to depend on the face of the organic crystal and thus the relative orientation of the molecules with respect to the surface crystal surface [30, 31]. This occurs despite the fact that a single molecule (in the gas phase) has one and only one IE_{gas} and EA_{gas} (Similar to small metal clusters with multiple facets of different crystal orientations have only one well-defined work function [32, 33]).

2.1.4. Energy level alignment at interfaces between conductive surfaces and conjugated organic molecules

Intimate contact between conductive (electrode) surfaces and conjugated organic molecules will result in a variety of effects occurring at the interface. On a most general level, one can state that upon contact formation the electron density distributions of both electrode and molecule change. This rearrangement is accompanied by changes of bond lengths and thus a structural relaxation of the molecule and/or the metal surface atoms may take place. Depending on the strength of the electronic rearrangement, the interaction is often classified between the two extreme cases of physisorption (weak) and chemisorption (strong). Besides the specific materials brought into contact to form the interface, also the preparation method (i.e. evaporation order, temperature, pressure, etc.) is important for the interaction strength.

Physisorption vs. chemisorption

The term physisorption (physical adsorption) describes a system in which the interaction is mainly mediated by van-der-Waals forces. The interaction can be well described by the Lennard-Jones potential [34]:

$$V(r) = 4V_0 \left[\left(\frac{d_{eff}}{r} \right)^{12} - \left(\frac{d_{eff}}{r} \right)^6 \right] \quad (2.3)$$

where V_0 is the minimal potential energy, d_{eff} the effective atomic/molecular diameter and r the distance between the components. The first term is due to the short range repulsive forces, which occur because of the Pauli exclusion principle and the Coulomb repulsion, while the second term includes the long range attractive van-der-Waals forces due to induced dipoles. In the case of molecular adsorption on surfaces, this means that the molecular and electronic structure remains almost undisturbed and that the bonding to the surface is rather weak. The energy gain upon adsorption of a physisorbed molecule is in the order of several kJ per mole [34]. In general, purely physisorbed atoms or molecules on surfaces will not be found because a weak electron density rearrangement will always occur (e.g. energy level broadening or "push-back" effect).

In chemisorbed (chemical adsorption) systems, strong interaction between the constituting atoms or molecules and the surface occurs. Depending on the extend of electron transfer, covalent or ionic bonds are formed between the atoms or molecules and the surface. Besides, the molecular conformation can be strongly influenced by the adsorption [35, 36]. The energy gain upon chemisorption can be higher by one order of magnitude compared to physisorbed systems [34].

The key question thus is how the different electron density rearrangements at electrode/molecule interfaces influence the energy level alignment, i.e. the positions of molecular energy levels with respect to the electrode Fermi level, and so Δ_h and Δ_e. Up to date no unified model for the description and prediction of the energy level alignment at the electrode/molecule interface exists. Nevertheless, several models have been proposed that are able to successfully describe certain subsets of interface types. In principle, Δ_h and Δ_e could be estimated using the substrate work function Φ_S and the (orientation dependent) IE and EA of the molecular film. However, it is known that an interface dipole ID across the newly formed interface can occur and Δ_h and Δ_e then become:

$$\Delta_h = IE - \Phi_S + ID \qquad (2.4)$$

$$\Delta_e = EA - \Phi_S + ID \qquad (2.5)$$

Vacuum level alignment

In the limiting case of vacuum-level alignment (Schottky-Mott limit, interface essential free of gap states) the ID is zero and thus the injection barriers can be directly obtained from separately determined IE, EA, and Φ_S. In this case the Δ_h (Δ_e) can be decreased easily by increasing (decreasing) the metal work function for a given organic material as shown in Fig. 2.4a and b. In general however, vacuum-level alignment is rather the exception than the rule at most interfaces with conjugated organic materials. This is particularly relevant for interfaces with clean metal surfaces, because an ID results almost in every case from the "push back" of electron density into the metal by the organic adsorbate [37, 38].

Chapter 2. Fundamentals of metal/organic interfaces

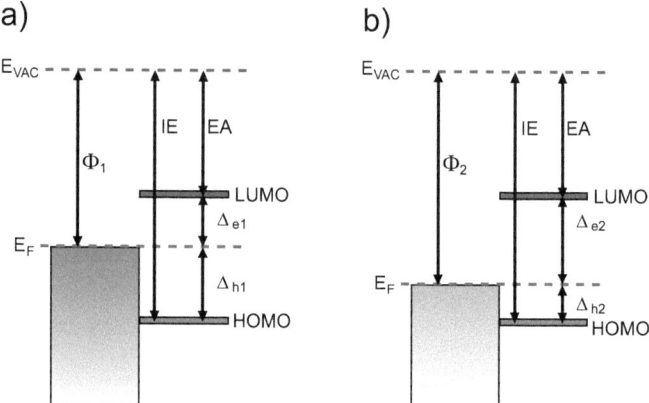

Figure 2.4.: Schematic energy level diagram for the case of vacuum-level alignment (Schottky-Mott-limit) between an organic layer adsorbed on an electrode surface. a) low work function (Φ_1) and b) a high work function (Φ_2) electrode. The corresponding changes of Δ_h and Δ_e are indicated. E_F denotes the Fermi level and E_{VAC} the vacuum level.

The "push-back" effect

The adsorption of noble gases (i.e. Xe) on clean metal surfaces can be regarded very close to physisorption, because in this limiting case, there is a negligible hybridization between the electronic levels of the adsorbate and the metal substrate [39]. Despite the weak interaction, the work function is lowered significantly upon adsorption because of the Coulomb repulsion between the electron density of the molecule and the surface electrons [20, 39, 40]. This locally "pushes" the tail of the electron wave function back into the metal and effectively decreases the SD and thus Φ. This is illustrated in Fig. 2.5a and b which stress that no new dipole is created, but the existing SD is changed. A second adsorbate that is close to physisorption on clean metal surfaces is tetratetra-contane (TTC). Here, almost no hybridization between molecule and metal bands occurs as TTC is a wide bandgap σ–insulator [38]. TTC reduces the work function upon adsorption on clean Au(111) surfaces by −0.7 eV, on Cu(111) by −0.3 eV to −0.5 eV, on Ag by −0.5 eV, and on Pb by −0.3 eV [41, 42, 43]. Similar work function reduction has been reported for initially clean metal surfaces that were intentionally contaminated by exposure to air or common solvents [44, 45]. Comparing the work function changes found for hydrocarbons with the changes upon adsorption of Xe on various metals, a similar dependence is found in both cases: the larger the initial work function of the metal, the larger the Φ reduction will be due to the decreased SD. The "push back" effect is thus a

2.1. Electronic structure of metal/organic interfaces

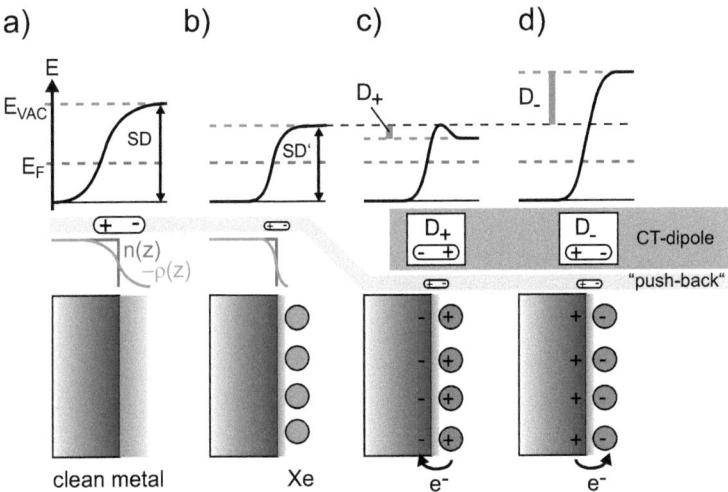

Figure 2.5.: Illustration of the effect of different adsorbates on Φ of metal surfaces: a) clean metal surface, b) xenon adsorption for a system close to physisorption, c) adsorption of molecular donor materials with net electron transfer from the adsorbate to the metal, d) adsorption of molecular acceptor materials with net electron transfer from the metal to the adsorbate. For a) and b) the changes in the electron tail $\rho(z)$ upon adsorption are indicated. The changes in electrostatic potential $\varphi(z)$ are shown in all cases, which defines the final vacuum level E_{VAC} and the work function Φ_S. D_+ and D_- indicate the dipoles originating from the charge transfer.

general phenomenon occurring upon adsorption of atoms or molecules on clean metal surfaces, with its magnitude being dependent on the substrate metals SD, the adsorption distance and effective surface coverage [37, 38, 46].

Fermi level pinning

In several cases vacuum level alignment can actually be observed for electrode/COM interfaces in a certain range of initial substrate work functions Φ_S. Then the change in Δ_h and Δ_e directly follows the change of the work function as shown in Fig. 2.6b. However, when Φ_S is decreased below or increased above a certain threshold value and thus comes close to the molecular IE or EA, pinning of molecular levels at the Fermi level occurs. This situation is depicted in Fig. 2.6a and c. If the work function is decreased below the threshold, a pinning of a molecular level P1 at the Fermi level occurs. Consequently, an interface dipole builds up

Chapter 2. Fundamentals of metal/organic interfaces

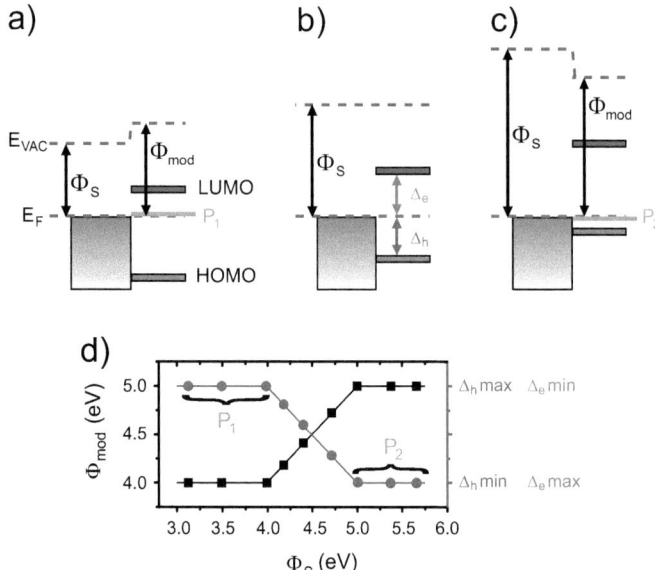

Figure 2.6.: Schematic energy level diagrams showing the two pinning regimes of Fermi level pinning: a) pinning of the LUMO with a level P1 at the Fermi level and c) pinning of the HOMO with a level P2 at the Fermi level. b) shows the intermediate region, where vacuum-level alignment prevails in cases where the electron "push-back" is absent. d) Plot of the adsorbate modified work function Φ_{mod} and the hole injection barrier Δ_h versus the initial substrate work function Φ_S.

due to the charge that must be transferred to reach thermal equilibrium. Further decrease of the work function will not change the position of the molecular levels, such as HOMO and LUMO, with respect to the Fermi level. In this case the injection barriers become independent of the substrate work function Φ_S. Thus, a maximum value for the Δ_h is reached, as shown in Fig. 2.6d, which corresponds to a minimum in Δ_e. In the case of high initial substrate work functions a pinning of a molecular level P2 at the Fermi level occurs, as shown in Fig. 2.6c. In this regime, a minimum value for the Δ_h is reached, as shown in Fig. 2.6d, corresponding to a maximum in Δ_e. Several models aiming to explain this behavior exist, the most prominent ones being the tail gap state model [47] and the integer charge transfer model [48, 49]. Generally, it is observed that Fermi level pinning occurs, when the interaction between the conductive substrate and the subsequently deposited organic molecules or polymers is not too

2.1. Electronic structure of metal/organic interfaces

strong, i.e. a negligible hybridization between molecular and substrate levels occurs. This is usually the case for oxidized and/or ambient air exposed metal and semiconductor surfaces, and generally also for clean metal surfaces onto which a thin layer of organic molecules has already been deposited. Furthermore, the "push back" effect can no longer occur at these passivated interfaces because it is already saturated due to the presence of the contaminants or molecules/oxides at the interface to the substrate. On clean metal surfaces, pinning of molecular levels at the Fermi level can also occur, here however, the interaction can be much stronger compared to the previous cases and vacuum-level alignment will probably not be observed in the region between the two pinning regimes as the "push-back" effect will be always present as discussed above.

Bond formation and charge transfer

When the interaction strength between substrate and molecule is increased, partial or integer charge transfer and/or the formation of covalent or ionic bonds occurs. The latter are usually formed when reactive metals are used in the interface formation [37].

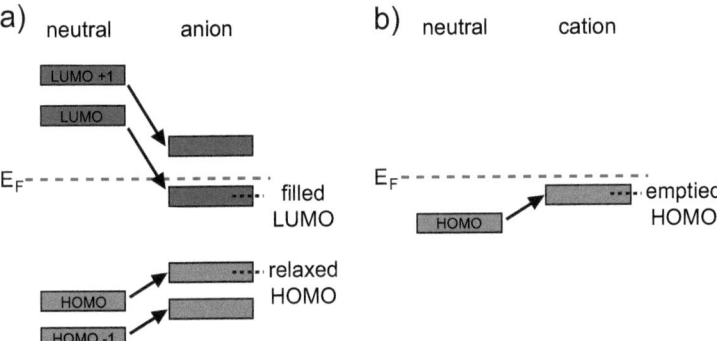

Figure 2.7.: Schematic energy level diagrams of the frontier orbitals of a) electron acceptor and b) electron donor molecules in the neutral and ionic state. As a reference the Fermi level is shown in both cases. The arrows indicate the origin of the newly distributed molecular levels in the charged donor/acceptor.

Charge transfer is generally observed for the adsorption of strong electron acceptor or donor materials on various surfaces. This is usually accompanied by the formation of hybrid states between the metal continuum states and specific molecular levels. In the case of strong electron acceptors, where the charge transfer (CT) is directed from the metal to the molecule, a metal-molecule hybrid state derived from the former LUMO of the neutral molecule gets (partially) filled and it thus located below the Fermi level in the chemisorbed layer (see

15

Chapter 2. Fundamentals of metal/organic interfaces

Fig. 2.7a). Since it is in many cases mainly located on the molecule, it is also often referred to as a simple filling of the former LUMO. The charge transfer also causes the molecular structure to relax into a new geometry [50, 16], which leads to new energetic positions of already filled molecular orbitals (e.g. HOMO). This is also indicated in Fig. 2.7a. The work function of the system is usually observed to be increased, because a dipole with its negative part pointing away from the surfaces is established across the interface by the charge transfer (D_- in Fig. 2.5d). On the other hand, it is also conceivable, that the charge transfer dipole may be affected by the decrease in work function due to the "push-back" effect. Then only a rather small Φ increase or even a decrease would be observed after adsorption. The latter has been observed for iodine on metal surfaces and is theoretically well understood [40]. In contrast charges are transferred from the molecule to the metal in the case of electron donor materials. In this case no widely accepted model for the energetic change in the frontier energy levels exists. However, it has been speculated that as the molecular orbitals lose electron density, all molecular orbitals should shift towards the Fermi level as shown in Fig. 2.7 [10, 11]. The work function is usually decreased beyond typical values of the "push-back" effect on these surfaces, because the charge transfer dipole (D_+) points in the same direction as shown in Fig. 2.5c.

2.1.5. Charge injection

In pure organic semiconductors, which are not subjected to light, electron or ion radiation (which could produce free charge carriers), charges can only flow if they are injected from an external source, because the large band gap (oligoacenes have $E_{G,\text{gas}}$ of several eV [21]) compared to the thermal energy at room temperature ($k_B T \approx 25$ meV) means that there are almost no intrinsic charge carriers available. Thus, as already discussed in the introduction, charge injection is crucial for the performance of organic electronic devices. In most cases the charge carriers are injected from metal electrodes into the organic semiconductor layers. Two limiting cases of injection across a metal semiconductor interfaces can be discerned. In the first case, the injection of charge carriers is limited by the supply of charges to the organic semiconductor layer. Here the charge carriers (the discussion will be restricted to electrons, however, all principles also apply for holes) need to overcome a certain barrier Δ_e in order to be injected into the organic semiconductor layer. This can either occur if the electron has enough kinetic energy due to thermal activation to overcome the barrier (thermionic emission) or by tunneling through the barrier (field emission). It can be easily understood, that thermionic emission will be dominant at high temperatures, whereas field emission will be the prevalent process at low temperatures and with high barriers. In both cases the Schottky effect (attraction of the image charge) plays an important role, as it reduces the barrier (as shown in Fig. 2.8) by $\delta_e = ((e^3 F)/(4\pi\epsilon\epsilon_0))^{(1/2)}$ [51, 21] with the elementary charge

2.1. Electronic structure of metal/organic interfaces

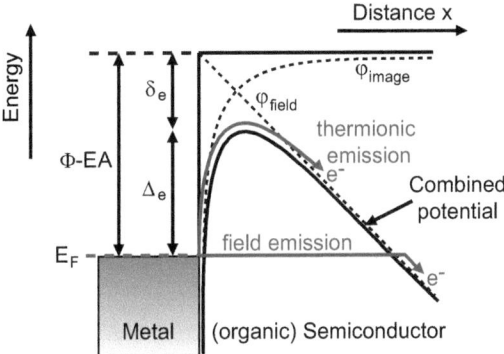

Figure 2.8.: Charge injection at a metal-organic interface. The two extreme cases of thermionic and field emission of the electron are shown by the red lines. The initial barrier height at the interface is the difference between the work function of the metal Φ and the electron affinity EA of the organic film. The potential of the applied electric field φ_{field} and the electrons image in the metal φ_{image} are plotted as dashed lines and the resulting combined potential is also indicated. The initial barrier height for pure thermionic emission is reduced due to these potentials by δ_e. The resulting electron injection barrier under applied bias for pure thermionic emission is thus δ_e. Field emission (tunneling through the barrier) presents the other pathway for charge injection.

e, the electric field F, the relative permittivity of the material ϵ, and the vacuum permittivity ϵ_0. The barrier reduction δ_e thus strongly depends on the applied electric field F. For a 500 nm thick organic film over which a voltage of 4 V is applied, which are typical value for light emitting diodes, δ_e is ≈ 0.03 eV. The superposition of the applied electric field $\varphi_{field} = -eF \cdot x$ and the image potential $\varphi_{image} = e^2/(16\pi\epsilon\epsilon_0) \cdot (1/x)$ creates a quasi triangular barrier with the height Δ_e as shown in Fig. 2.8 from a step like barrier of height $\Phi - EA$.

The current density j_{RS} including the image charge potential for thermionic emission is given by the Richardson formula [52, 21]:

$$j_{RS} = AT^2 \exp\left(\frac{-\Phi_B}{k_B T}\right) \quad (2.6)$$

with the Richardson constant $A = 4\pi em^*(k^2/h^3)$ (m^* is the reduced mass of the charge carriers) and the absolute temperature T. The current density j_{FN} for field emission, as the other limiting case, through a triangular barrier is described by the Fowler-Nordheim formula

Chapter 2. Fundamentals of metal/organic interfaces

[52, 21]:

$$j_{FN} = \frac{A}{\Phi_B} \left(\frac{eF}{\alpha k_B}\right)^2 \exp\left(\frac{-2\alpha \Phi_B^{\frac{3}{2}}}{3eF}\right) \qquad (2.7)$$

where $\alpha = 4\pi(\sqrt{2m^*}/h)$. In an ideal Schottky-diode both effects can occur separately or in a combined process called thermionic-field emission. The current density j_S through an ideal Schottky-diode is thus described by the Shockley formula [52, 21]:

$$j_S = AT^2 \exp\left(\frac{-\Phi_B}{k_B T}\right) \exp\left(\frac{eU_{ext}}{\beta k_B T} - 1\right) \qquad (2.8)$$

with β as an "ideality" factor of the Schottky-diode and U_{ext} as the externally applied bias. This shows the exponential dependence of j_S on the barrier height Φ_B with the necessity of low injection barriers for electrons (and holes) as a small decrease of 50 meV in the barrier will already increase the current by one order of magnitude.

If the current is limited by the intrinsic free charge carriers of the organic semiconductor, the current is generally termed space-charge-limited-current (SCLC). Sometimes the contacts in this regime are termed ohmic. This is, however, misleading, because it does not necessarily mean that the current over the interface is ohmic. Only for small voltages U_{ext}, where fewer charges are injected than are intrinsically available per volume (N_0), the current density j_{OHM} across the interface will follow Ohm's law:

$$j_{OHM} = eN_0 \mu \frac{U}{L} \qquad (2.9)$$

where μ is the mobility of the intrinsically available charge carriers in the organic material, and L the sample length in direction of the current flow. As soon as the applied voltage increases such that there are more charge carriers injected than available intrinsically in the organic material, a space-charge will build up at the interface in the organic semiconductor and current density will increase super-linear with the applied voltage U_{ext}. If there exist no traps, which could originate from impurities or defects, then the current density $j_{TF-SCLC}$ is denoted as trap free space-charge-limited and given by the Mott-Gurney formula [52]:

$$j_{SCLC} = \frac{9}{8} \epsilon \epsilon_0 \mu^{COM} \frac{U^2}{L^3} \qquad (2.10)$$

However, in general shallow traps with various energies and deeper discrete traps are present in organic semiconductors. From shallow traps captured charge carriers can be thermally re-activated. From deep traps this is not possible [21]. The shallow traps will be filled first and thus the mobility is modified to an effective mobility $\mu^{COM} \rightarrow \mu_{eff}^{COM} = \theta_{tr} \mu^{COM}$ [21] with θ_{tr} defined by the traps. After also the discrete traps are filled, $\theta \approx 1$, and the current density will be just as large as without the traps in this so called trap-filled limit (TFL). At the voltage U_{TFL} the current density will strongly increase as shown in Fig. 2.9.

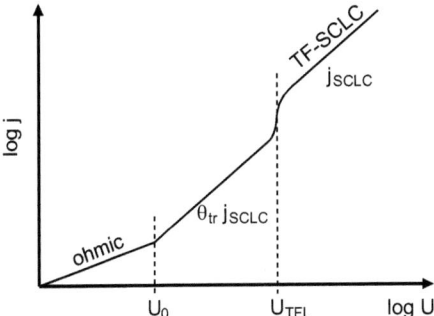

Figure 2.9.: The current density versus voltage characteristics through an organic semiconductor with ohmic contacts exhibiting ohmic and space charge limited current. U_0 denotes the threshold voltage between ohmic and SCLC behavior, U_{TFL} the critical voltage of trap filling and θ_{tr} is a parameter accounting for the density and depth of the traps.

2.2. Morphological structure of metal/organic interfaces

2.2.1. Growth

Organic thin films are usually grown on various substrates under vacuum conditions by molecular beam deposition from resistively heated sources. The growth or more precise: the growth mode and orientation of the organic material deposited on various substrates can strongly influence the electronic structure (e.g. the overlap between molecular orbitals and substrate bands, the ionization energy as discussed in Sec. 2.1.3, etc.). During growth, a certain number of molecules per time interval will reach the surface and, depending on the sticking coefficient between molecules and surface, stay there. When the molecule is adsorbed on the surface, a variety of dynamic processes can occur as shown in Fig. 2.10. Possible processes include inter- and intra-layer diffusion, nucleation, dissociation, and adsorption at special sites, such as step edges, defects, etc. [53]. The nucleation of molecules will be stable if a certain number of molecules meet at the surface, which is called the critical size of the nucleus. If the number is smaller than necessary dissociation will occur. The growth of the organic material is generally far from equilibrium and thus a non-equilibrium kinetic process, where the final film morphology depends on the route taken through the processes described in Fig. 2.10. Therefore, the final film morphology is not necessarily the most stable one as shown in sev-

Chapter 2. Fundamentals of metal/organic interfaces

eral experiments, where grown films were annealed and a change in morphology was observed [54, 55]. To describe the dynamic processes of growth several models exist [53, 56, 57, 58], that have been applied to the growth of organic materials [58, 59, 60, 61].

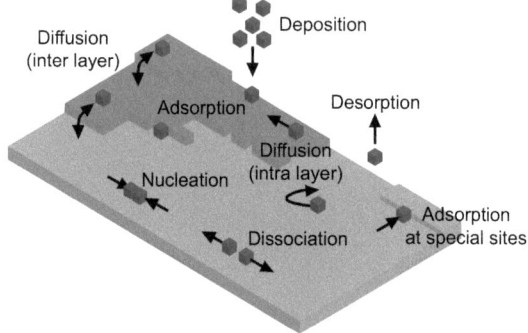

Figure 2.10.: Model surface indicating the different processes occurring upon deposition of an organic material.

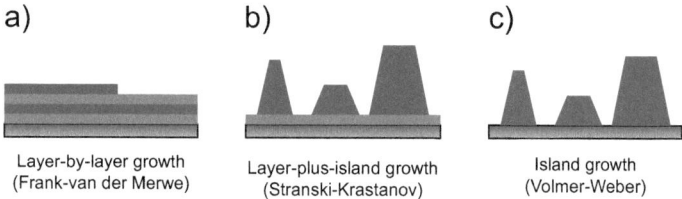

Figure 2.11.: Different growth modes observed for (organic) materials deposited on solid substrates.

In general the film morphology depends critically on the relative strength of the molecule-substrate and molecule-molecule interactions, which leads to three main growth modes as shown in Fig. 2.11. If the interaction between the substrate and the molecules is stronger than the intermolecular interaction, layer-by-layer growth will result (Frank-van der Merwe growth). In contrast, if the intermolecular interactions are exceeding the molecule-substrate interaction, island growth is found starting already with the first layer (Volmer-Weber growth). An intermediate case, where islands nucleate on a (few) closed monolayer(s) is often found in organic thin film growth (Stranski-Krastanov growth), where subsequent layer growth becomes unfavorable at a certain film thickness. Compared to inorganic thin film growth, the most fundamental difference to organic thin film growth lies in the fact, that

2.2. Morphological structure of metal/organic interfaces

organic molecules are extended objects and thus have internal degrees of freedom. Consequently, the orientation of the molecules has to be included in the growth models. In general, the literature distinguishes between two cases shown in Fig. 2.12, where again the relative strength between molecule-substrate and molecule-molecule interactions play a crucial role. In the first case, where the molecule-substrate interaction prevails, the molecules in the first molecular layer tend to maximize the overlap of their π-system with the metal bands and thus adopt a face-on conformation (Fig. 2.12a). This is found for π-conjugated molecules on clean metal surfaces [50, 55, 62, 63, 64, 65, 66, 67, 68, 69]. Only multilayers are known to eventually adopt a different orientation [55, 65, 70] because the strong interaction with the metal is already screened by the monolayer. In the second case, where the molecule-molecule dominate over the molecule-substrate interactions, it is found that the molecules try to maximize their π-system-overlap and minimize their surface footprint. Consequently, they adopt an edge-on conformation (Fig. 2.12b), in general with a certain tilt angle θ_t between long molecular axis and surface normal. Examples are various kinds of oxides, such as ITO, TiO_2, SiO_x, etc. [31, 60, 71, 72, 73, 74, 75] and ambient exposed metal surfaces [45, 47] or self assembled monolayer (SAM) covered surfaces [76, 77]. In general the edge-on orientation of the monolayer is retained in further film growth on these passivated surfaces.

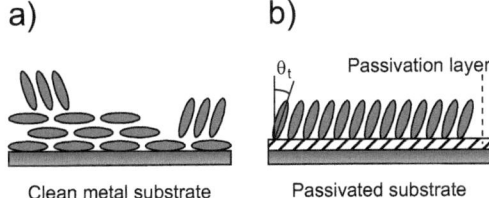

Figure 2.12.: Orientation dependent growth of organic materials on a) clean metal and b) passivated surfaces. In the first case the metal-molecule and in the latter the molecule-molecule interactions prevail.

Chapter 3.

Experimental methods

In this chapter the experimental methods used in this work to characterize the molecule and metal systems are introduced. Mainly photoelectron and reflection absorption infrared spectroscopies were used for which the theoretical background shall be given here. The experimental setups, which were used for the experiments will be illustrated in the next chapter.

3.1. Photoelectron spectroscopy

In the past decades photoelectron spectroscopy has been and is still widely used for studying the electronic and chemical properties of materials. The working principle is based on the external photoelectric effect [78, 79], where electrons are excited by an incident light beam with (monochromatic) energy $h\nu$ and then emitted from the sample if their kinetic energy E_{kin} is larger than the work function of the material Φ_S:

$$E_{kin} = h\nu - \Phi_S \qquad (3.1)$$

PES can be subdivided depending on the excitation photon energies used: ultraviolet (energy range: 10 eV to 100 eV) and X-ray (100 eV and above) photoelectron spectroscopy - UPS and XPS, respectively. In both cases PES is a surface sensitive technique, because the inelastic mean-free path λ of the electrons in solids is rather short as obtained from the universal mean-free path curve (2-10 Å for UPS and up to a few tens of Å for XPS) [80]. Therefore, PES is the technique of choice for surface, interface, and thin film studies. Note that PES is generally a surface averaging technique because the analyzed sample spot size is usually in the range of μm^2 or even mm^2. To obtain reliable data, sample charging and photo-degradation need to be avoided, which is a particularly important issue for organic materials. Photoelectrons leave the sample ionized and charge neutrality must be re-established through current from the substrate. This may lead to positive charge build-up at the sample surface because of the low conductivity of many organic materials and subsequent chemical reactions may happen

Chapter 3. Experimental methods

between positively charged molecules [81]. Therefore, in all experiments great care must be taken with regard to this issue. Eventual charging can often be efficiently removed by sample illumination with visible light and thus generation of additional mobile charges via internal photoemission at the interface to the substrate [82, 83].

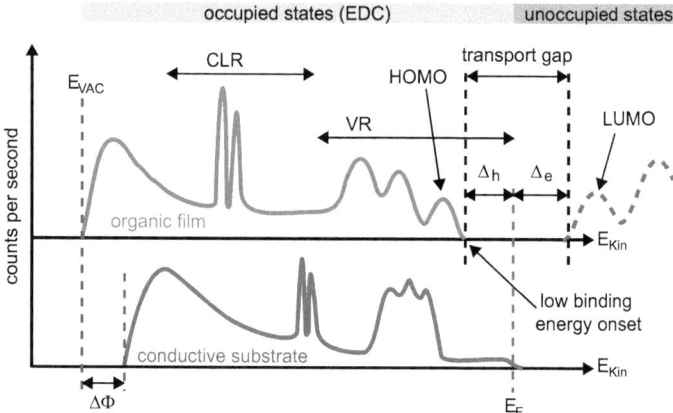

Figure 3.1.: Energy distribution curve (EDC) obtained in a typical UPS experiment for a conductive substrate and a conjugated organic film (occupied states). In addition, unoccupied molecular states accessible using inverse photoemission techniques are also indicated. The valence (VR) and core level region (CLR) are indicated together with the hole and electron injection barrier (Δ_h and Δ_e), which is defined as the energetic difference between the Fermi level (E_F) and the onset of the highest occupied and lowest unoccupied molecular orbital (HOMO and LUMO respectively) of the organic film. The local vacuum level is indicated by E_{VAC}, which permits calculation of the sample work function. The change in work function between metallic and organic film is denoted $\Delta\Phi$.

A typical energy distribution curve (EDC) obtained in a PES experiment is shown in Fig. 3.1, where the number of counted photoelectrons are plotted against their kinetic energy. The electrons with highest kinetic energy E_{kin} stem from the highest occupied states of the sample, which is the Fermi level (E_F) for a metal or the HOMO for an organic layer. Commonly, E_F is the reference level in PES, because it is aligned between sample and analyzer as they are electrically connected and in thermal equilibrium. Moving in the spectrum in Fig. 3.1 to slightly lower E_{kin}, rather broad peaks or structured features are observed that stem from the valence region (VR) of the sample. To first order approximation (i.e., without accounting for cross-section and selection rule effects, which will be discussed in Sec. 3.1.1), PES yields the density of occupied levels of a solid. The VR photoelectrons are thus related

3.1. Photoelectron spectroscopy

to valence bands of extended systems (covalent or van-der-Waals crystals, or polymers) or various σ- and π-type orbitals of molecules. The measured electrons in the core level region (CLR) in Fig. 3.1 result from the excitation of deep lying localized atomic orbitals that are well shielded by the valence electrons. At very low kinetic energies the main contribution to the spectrum are electrons that have undergone one or more inelastic scattering processes in the sample, i.e., secondary electrons that have lost the information about their initial state. The sharp low-energy cut off (often called secondary electron cut off - SECO) is a direct measure of the local vacuum level (E_{VAC}) position (in front of the sample, in contrast to the vacuum level at infinity $\Phi_\infty{}^1$), as the electrons have just enough energy to overcome Φ_S and leave the sample in the limit $E_{kin} \to 0$ eV (see Eq. 3.1).

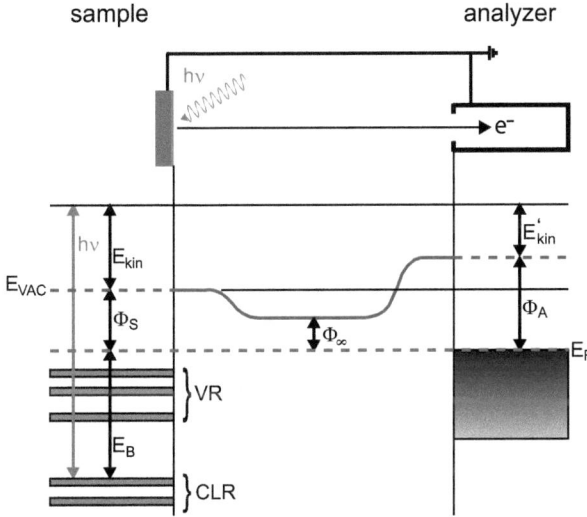

Figure 3.2.: Schematic energy level diagram for the sample and the electron energy analyzer in a photoelectron spectroscopy (PES) experiment. The upper part presents the principal experimental setup consisting of a sample and the electron energy analyzer. E_F denotes the common Fermi level, E_{VAC} the vacuum level, $h\nu$ the photon energy, Φ_S and Φ_A the work functions of the sample and analyzer, E_{kin} and E'_{kin} the kinetic energy of the electrons above the sample and the analyzer, Φ_∞ the work function at infinity, E_B the binding energy of the electrons in the sample, and VR the valence and CLR the core level region of occupied states in the sample.

---

[1] See Ref. [84] for a detailed discussion.

Chapter 3. Experimental methods

However, in a real experiment the work function of the analyzer Φ_A needs to be considered as well, which is drawn to be larger than Φ_S in Fig. 3.2; the reverse situation may also occur. Thus, all electrons with $E_{kin} < \Phi_A - \Phi_S$ are unable to reach the analyzer. Consequently, a constant negative potential U_{SECO} is applied to the sample when measuring the SECO (typically a few eV). This rigidly shifts the whole EDC to higher E_{kin} by the constant value of U_{SECO} and ensures that all electrons are indeed detected. The difference between Φ_S and Φ_A is also responsible for the variation of E_{VAC} along the electron path as shown in Fig. 3.2. However, the measured kinetic energy E'_{kin} is unaffected by this variation, because only the potential difference between the sample surface and the spectrometer is relevant (as electrostatic interaction is conservative). Consequently the following relation applies for E'_{kin}:

$$E'_{kin} = h\nu - E_B - \Phi_A = E_{kin} - (\Phi_A - \Phi_S) \tag{3.2}$$

where E_B denotes the binding energy with respect to E_F. In an experiment Φ_A is constant and thus E_B can be measured. Φ_S can be obtained by subtracting the full width of the EDC (with applied bias U_{SECO} when measuring $E_{kin,SECO}$) from the photon energy:

$$\Phi_S = h\nu - (E_{kin,E_F} - E_{kin,SECO} - U_{SECO}) \tag{3.3}$$

The hole injection barrier (Δ_h) as important electronic device relevant interface parameter, which is (per definition) the difference between the Fermi level and the low binding energy onset (see Fig. 3.2 and 4.4) of the HOMO of the organic overlayers, can be extracted by:

$$\Delta_h = E_{kin,E_F} - E_{kin,HOMO\,onset} \tag{3.4}$$

Combining both values yields the ionization energy (IE) of the organic material, which is (per definition) the difference between E_{VAC} and $E_{kin,HOMO\,onset}$:

$$IE = h\nu - (E_{kin,HOMO\,onset} - E_{kin,SECO} - U_{SECO}) = \Phi_S + \Delta_h \tag{3.5}$$

In addition to the direct information from VR density of states distribution, the exact binding energies of core levels measured by XPS provide valuable information on the chemical environment and bonding configuration of specific atomic species. The change in binding energy between two different chemical forms of the same atom is called *chemical shift*. It arises from the fact, that neighboring chemical bonds determine the local electrostatic potential for the core levels of an atom. This can also be regarded as a many body effect, since different local electrostatic potentials influence the efficiency of the photo-core-hole shielding/screening and thus the measured kinetic energy of the escaping electron. The effects of different chemical environments on the core level binding energy is nicely illustrated in Fig. 3.3 for the C1s core level of ethyl-trifluoroacetate. Chemical shifts of up to 11 eV have been reported for carbon atoms in different chemical environments [85].

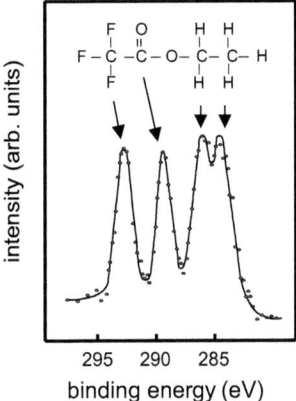

Figure 3.3.: Carbon core level spectrum of ethyl-trifluoroacetate showing the chemical shift of carbon atoms with different chemical environments. Adapted from Ref. [86].

3.1.1. Theoretical background

For the theoretical description of the photoemission process, as described in the previous section, the single particle picture shall be used here rather than the complex many-body treatment [87] (and references therein). Many-body effects necessary to understand the spectra presented in this work will be illustrated in Sec. 3.1.4. In the single-particle picture, the "sudden-approximation" [88] is used, which assumes that the removal of the electron from the remaining N-1 electron system is so fast, that there is only negligible interaction between the two systems during ejection[2]. To describe the photoemission process, the so called three-step model has proven to describe the basic spectral features quite well [87, 88, 89]. In this model, the photoemission process is divided into:

1. Photoexcitation of an electron in the solid.

2. Traveling of the exited electron to the surface of the solid.

3. The escape of the electron from the solid into the vacuum trough the surface.

---
[2]This is the case for metals, since there the shielding/screening of the photo-hole by the remaining electrons is almost perfect and thus the escaping electron remains undisturbed

Chapter 3. Experimental methods

Consequently, the total external emission current I^{ext} can be written as a product of factors which are determined individually in three steps [89, 90, 91]:

$$I^{ext}(E'_{kin}, h\nu, \vec{k}) = I^{int}(E_{kin}, h\nu, \vec{k}) \cdot T(E_{kin}) \cdot X(E_{kin}, \vec{k}) \quad (3.6)$$

where E_{kin} denotes the kinetic energy of the electron, $h\nu$ the photon energy, $I^{int}(E_{kin}, h\nu, \vec{k})$ the internal electron current, $T(E_{kin}, \vec{k})$ the transport function, and $X(E_{kin}, \vec{k})$ the escape function.

First step. The transition probability W per unit time for the photoexcitation of an electron from the initial state Ψ_i to a final state Ψ_f, can be obtained by Fermi's golden rule, if the perturbation H' is small:

$$W = \frac{2\pi}{\hbar} \mid \langle \Psi_f \mid H' \mid \Psi_i \rangle \mid^2 \delta\left(E_f(\vec{k}) - E_i(\vec{k}) - h\nu\right) \quad (3.7)$$

with $E_f(\vec{k})$ and $E_i(\vec{k})$ being the energies of the final and initial states. In Eq. 3.7, the δ-function ensures the energy conservation. The momentum conservation is hidden in the first factor, because H contains the potential $V(\vec{r})$ [91]. The perturbation H' can be written as:

$$H' = \frac{e}{2m_e c}\left(\vec{A} \cdot \underline{P} + \underline{P} \cdot \vec{A}\right) - e\Phi_{light} + \frac{e^2}{2m_e c^2}\vec{A} \cdot \vec{A} \quad (3.8)$$

where e and m_e are the charge and mass of the electron, c the velocity of light, $\vec{A}$ and Φ_{light} are the vector and scalar potential of the incident light, and $\underline{P}$ the momentum operator. Usually now several approximations are being made to Eq. 3.8 [88, 91]: First of all, the term $\vec{A} \cdot \vec{A}$ is neglected, since it does not contribute to a one-photon process. Second, Φ_{light} becomes zero in the Coulomb gauge. Third, the wavelength of the incident light is large compared to atomic distances, which permits the use of the dipole approximation for the vector potential: $\vec{A} = \vec{A}_0 \exp(i\vec{q} \cdot \vec{r}) \approx \vec{A}_0$ and the assumption that $\vec{A}_0$ and $\underline{P}$ commute. Thus the perturbation H' becomes:

$$H' = \frac{e}{m_e c}\left(\vec{A}_0 \cdot \underline{P}\right) \quad (3.9)$$

Then the transition probability W can be written as:

$$W = \frac{2\pi}{\hbar}\frac{e}{2m_e c} \mid \langle \Psi_f \mid \vec{A}_0 \cdot \underline{P} \mid \Psi_i \rangle \mid^2 \delta\left(E_f(\vec{k}) - E_i(\vec{k}) - h\nu\right) \quad (3.10)$$

with the transition dipole moment $M_{i \to f} = \mid \langle \Psi_f \mid \vec{A}_0 \cdot \underline{P} \mid \Psi_i \rangle \mid^2$. Consequently, the internal electron current density $I^{int}(E_{kin}, h\nu, \vec{k})$ generated by the optical excitation is

3.1. Photoelectron spectroscopy

obtained by summation over all initial and final states i and f [91]:

$$I^{int}(E_{kin}, h\nu, \vec{k}) \propto \sum_{f,i} M_{i\to f} \delta\left(E_f(\vec{k}) - E_i(\vec{k}) - h\nu\right) \delta\left(E_{kin} - E_f(\vec{k})\right) \quad (3.11)$$

where $\vec{k}$ needs to be directed towards the surface. To detect an electron at the energy E_{kin}, the energy of the final state $E_f(\vec{k})$ needs to equal E_{kin} (as stated before this is adjusted by the energy analyzer) and E_{kin} must be large enough to overcome the work function of the sample Φ_S. From the first δ-function in Eq. 3.11 it can be seen, that the electron current density is proportional to the density of occupied initial states in the solid. Thus, the EDC in an PES experiment is, within the approximations of the three-step model, a mapping of the density of occupied states (below the Fermi level).

If the energy and momentum conservation in Eq. 3.11 is met, then the transition dipole moment $M_{i\to f}$ needs to be non-vanishing for a certain initial Ψ_i and final state Ψ_f. If this is the case, then the transition will be allowed. This presents the *dipole selection rule of photoemission spectroscopy*. The condition for a non-vanishing transition dipole moment is that the direct product of the irreducible representations[3] [93, 92] of Ψ_i, $\vec{A}_0 \cdot \underline{P}$, and Ψ_f contains the totally symmetric irreducible representation (in general A_1), as can be derived from group theory. Further details for the symmetry of the initial state and the selection rule will be given later (Sec. 3.1.3).

Second step. In this step the electrons described within Eq. 3.11 propagate to the surface of the solid. Some of these electrons will undergo one or more inelastic scattering processes on the way to the surface and loose a part of their energy E and all information about their initial state. They will contribute to the secondary electron background in the spectrum, if their energy at the surface is large enough to overcome Φ_S. The probability that an electron will reach the surface without inelastic scattering can be described by the mean-free electron path $\lambda(E_{kin})$, which can be derived from the universal mean-free-path curve [80] as stated earlier. As noted before, this step makes photoelectron spectroscopy a surface sensitive technique. Thus, the propagation can be described simply by a transport function $T(E_{kin})$ which is proportional to $\lambda(E_{kin})$:

$$T(E_{kin}) \propto \lambda(E_{kin}) \quad (3.12)$$

Third step. In the last step, the transmission of the electron from the solid to the vacuum

---
[3]The sets of characters for all symmetry operations of a point group are referred to as irreducible representations of the particular group [92]. They indicate how a molecular property (translation, operators, etc.) transforms under all symmetry operations of the point group. They are listed for the point groups important for this work in the character tables in the appendix A.1

Chapter 3. Experimental methods

through the surface needs to be considered. In this process every final state Ψ_f can be described by a Bloch electron wave, which is scattered by a surface-atom potential [91, 94]. The transmission of an electron through the surface into the vacuum requires conservation of its $\vec{k}$-vector component parallel to the surface, because of the 2D translational symmetry of the surface [94]:

$$\vec{k}_{\parallel}^{ext} = \vec{k}_{\parallel} + \vec{G} \qquad (3.13)$$

with $\vec{k}_{\parallel}^{ext}$ and $\vec{k}_{\parallel}$ being external and internal parallel component of the $\vec{k}$-vector and $\vec{G}_{\parallel}$ the parallel component of the reciprocal lattice vector. The normal component of the $\vec{k}$-vector is not conserved during the transmission. However, the normal component in the vacuum $k_{\perp}^{ext}$ can be obtained from the energy conservation:

$$E_{kin} = E_f - E_{VAC} = h\nu - E_B - \Phi_S = \frac{\hbar}{2m}(\vec{k}_{\parallel}^{ext2} + \vec{k}_{\perp}^{ext2}) \qquad (3.14)$$

where E_{kin} is the kinetic energy of the electron in the vacuum and E_B denotes the *positive* binding energy, with respect to the Fermi level. The parallel component $\vec{k}_{\parallel}^{ext}$ is linked to the experimental parameters with the following relation [94]:

$$\vec{k}_{\parallel}^{ext} = \sqrt{\frac{2m}{\hbar^2}E_{kin}}\sin(\alpha) \qquad (3.15)$$

where α is the angle under which the electron escapes the solid with respect to the surface normal. From Eqs. 3.13, 3.14, and 3.15 and by using parabolic final states of the form:

$$E_f = \frac{\hbar^2}{2m}(\vec{k}_f + \vec{G})^2 \qquad (3.16)$$

the equation for $k_{\perp}^{ext}$ can be obtained [94]:

$$\vec{k}_{\perp}^{ext} = \sqrt{\frac{2m}{\hbar^2}E_{kin}}\cos(\alpha) \qquad (3.17)$$

Consequently, the escape function $X(E_{kin}, \vec{k})$ depends not only on the energy E_{kin} of the final state, but via Eq. 3.13 also on the components of $\vec{k}$ in the vacuum. This can also be described via Eq. 3.15 using the angle α [4].

---

[4]In general also the azimuthal angle, which describes the rotation around the surface normal needs to be specified. However here, the description is solely given in the plane spanned between incident light and analyzer.

3.1. Photoelectron spectroscopy

3.1.2. Line width and shape

Not only the energetic position of a spectral feature can give valuable information, but also its shape. In general different broadening mechanisms have to be taken into account. Broadening induced by the experimental setup is due to the energetic width of the excitation source and instrumental contributions from the analyzer. Other sources are inhomogeneities of the sample and the lifetime broadening due to the excited state [94]. Inhomogeneous broadening can result from hole-vibrational coupling leading to the observation of vibrational progressions of the probed molecule at the high BE tail of a valence band peak [95].

3.1.3. Selection rules

The photo-induced transition from an initial Ψ_i to a final state Ψ_f in a PES experiment is determined by the transition dipole moment $M_{i \to f} = |\langle \Psi_f | \vec{A}_0 \cdot \underline{P} | \Psi_i \rangle|^2$. As stated before, the transition dipole moment must be non-vanishing in order for a transition to be allowed. This is the case if the direct product of Ψ_i, $\vec{A}_0 \cdot \underline{P}$, and Ψ_f in the integral contains the totally symmetric irreducible representation. For the general case it is impossible to evaluate the integrals $M_{i \to f}$. However, for some special cases with certain experimental geometries and symmetry of the electronic states involved, some predictions for the observation of the initial state can be made [94]. The assumptions made here for simplification of the problem are: relativistic effects are excluded, the surface has a mirror plane (yz) and $\vec{k}^{h\nu}$ of the incident light and the detection direction of the emitted electrons lie in this plane as shown in Fig. 3.4.

If a polarized light source is used (i.e. a dipole magnet at a synchrotron) the incident light can either be polarized parallel ($\vec{A}_1$) or perpendicular ($\vec{A}_2$) to the plane of incidence. In this work only light polarized parallel to the plane of incidence was used and thus only $\vec{A}_1$ should be considered in the following. In this case the product of $\vec{A}_1$ and $\underline{P}$ yields only y and z components, making this factor symmetric with respect to reflections in the mirror plane. Consequently, the remaining components of the transition dipole moment are:

$$M^y_{i \to f} = |\langle \Psi_f | \frac{\partial}{\partial y} | \Psi_i \rangle|^2 \qquad (3.18)$$

$$M^z_{i \to f} = |\langle \Psi_f | \frac{\partial}{\partial z} | \Psi_i \rangle|^2 \qquad (3.19)$$

The spatial components of the momentum operator ($\frac{\partial}{\partial y}, \frac{\partial}{\partial z}$) transform as the coordinates and thus the irreducible representations of the corresponding point groups can be looked up in the character tables in the appendix A.1. The initial state can either be a molecular orbital or a Bloch wave of an electron in the solid. The final state is an electron traveling in the plane of detection and must be symmetric/even with respect to this plane. In the case of

Chapter 3. Experimental methods

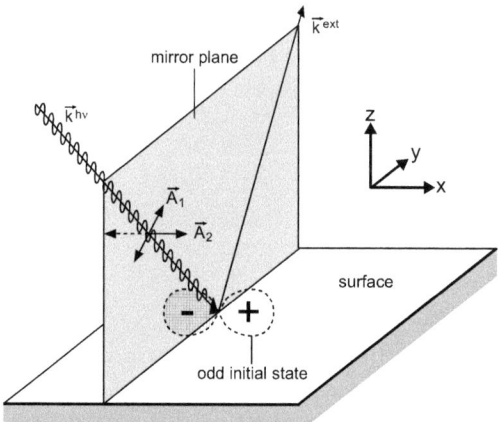

Figure 3.4.: Symmetry selection rule in a PES experiment. The direction of the incident light ($\vec{k}^{h\nu}$) and the trajectory of the photoemitted electron ($\vec{k}^{ext}$) lie in a mirror plane. The incident light is characterized by its polarization, which can be divided into two components $\vec{A}_1$ and $\vec{A}_2$. In the case shown here, the initial state from which the electron is emitted is antisymmetric with respect to the mirror plane. Only the component $\vec{A}_2$ gives rise to a measurable intensity at the detector if the final state is even with respect to the mirror plane.

normal emission, the final state must be totally symmetric. The respective symmetries for the corresponding point group of the sample (i.e. molecule adsorbed on a surface) can be looked up in the character tables in the appendix A.1. A valuable tool for obtaining the symmetry of complex molecular orbitals is the program Gaussian (further described in Sec. 4.5), which lists the symmetries of all molecular orbitals depending on the molecular point group[5]. *For an electric dipole transition from an initial to a final state to be allowed, the direct product of the three symmetry representations must contain the totally symmetric representation.* Only then the transition dipole moment will have a non-zero value as does the photocurrent.

3.1.4. Many-body effects

In the previous section, the theory of PES was described using the single-particle picture and the "sudden-approximation". Thus it was assumed, that the removal of an electron left the remaining N-1 electron system unchanged. However, the interaction of the emitted electron

---

[5]For molecules adsorbed on surfaces, the mirror plane in the plane of the molecule is generally lost. In Gaussian it was then also possible to use this reduced symmetry for the calculations

3.1. Photoelectron spectroscopy

with other electrons and the remaining photo-hole can in general not be neglected.

After the electron is emitted, which occurs on a time scale of $\approx 10^{-15}$ s, the (N-1) electron system (the molecular ion) is left in an excited state. The electronic relaxation from this state to the ground state of the ion occurs on the same time scale as the emission process. The polarization of (or screening of the created hole by) the surrounding dielectric media (molecules and/or substrate) takes place also on a time scale as short as 10^{-15} s, and is thus included in a PES measurement. However, the relaxation of the nuclei from the N to the N-1 electron system is generally not included, and the Frank-Condon principle is valid for PES experiments [96]. The polarization of the surrounding dielectric media by the photo-hole depends critically on the polarizability of the media. For materials with high polarizability, like metals, the screening of the hole by the electron cloud is very efficient, which reduces the coulombic interaction between hole and escaping electron. If the hole is created in materials with comparably smaller polarizability, like molecular crystals, screening is less efficient. Consequently, the attractive interaction between hole and escaping electron is stronger and E_{kin} of the electron is reduced (E_B increased) compared to the former case. This is especially important for molecular films on metallic substrates, because for sub- and monolayer films the metal half-sphere, which is underneath the molecules, is mainly responsible for the screening as depicted in Fig. 3.5. With increasing distance from the metal this influence becomes less important until it vanishes for thick organic films (Fig. 3.5). The photoelectron kinetic energy difference due to screening between a molecular layer in direct contact with a metal and the molecular bulk is typically in the range of 0.1−0.4 eV [38, 97, 98]. This relatively wide spread of energy differences is owed to the fact that it is experimentally extremely difficult to differentiate the screening effect from changes of a molecule's energy spectrum due to molecular conformation changes in the metal-adsorbed state [97].

Other many-body effects are especially important for the evaluation of XPS spectra, since the probed states are localized core levels. One of them is the Auger process, in which the recombination of the photo-hole is accompanied by the emission of an electron with a characteristic energy, as will be discussed in further detail in Sec. 3.1.5. Consequently, the XPS spectra will show element-specific Auger lines besides the photo-electron lines. Additionally the XPS spectra can show satellite peaks at higher BE (lower E_{kin}) compared to the main photo-electron lines due to shake-up and shake-off processes. They result from two-electron processes in which the emission of a photo-electron is accompanied by the excitation of an electron (in most cases from the valence band) to a bound state E_{bound} (shake-up) or to an unbound continuum state (shake-off), as shown schematically in Fig. 3.6. The energy needed to excite the second electron is consequently removed from the initially emitted photo-electron. The energy threshold for shake-up processes is the lowest possible transition from a filled to an empty state. For molecular layers this is usually the HOMO-LUMO transition (generally

Chapter 3. Experimental methods

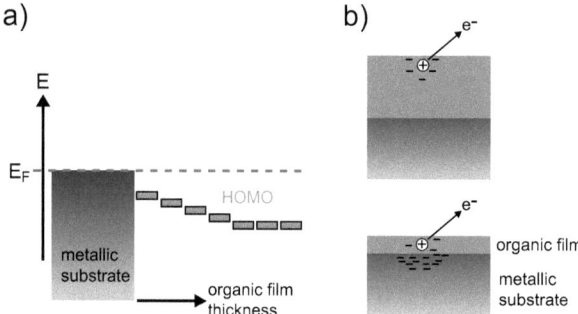

Figure 3.5.: a) Schematic energy level diagram of the HOMO position of an organic material as a function of film thickness. b) Schematic picture showing the effect of photo-hole screening for two different environments. Lower part: the photo hole is created in a monolayer film, upper part: the photo hole is created at the surface of a thick organic film.

in the order of some eV). In the case of metals or molecular layers that have density of states up to the Fermi level, the transition energy can be rather small and the shake-up peak will be very close to the main photo-electron line. In addition, multiplet splitting can occur due to the coupling of the electron spin with the momentum of its orbit (LS-coupling). This will result in a doublet splitting of the final state, with the intensities dependent on the degeneracy $g = 2 \cdot j + 1$ (with $j = l \pm s$) of the orbital.

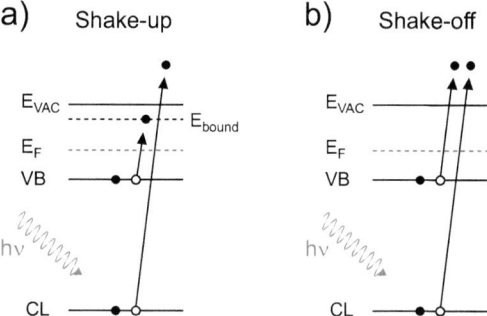

Figure 3.6.: Multi-electron processes occurring during core level photoexitation. a) shake-up and b) shake-off.

3.1. Photoelectron spectroscopy

3.1.5. Auger electron spectroscopy

An atom which has been ionized in one of its inner core levels (either by the radiation with light or electrons) can return to its ground state via an electron transition from an outer shell that fills the inner hole. This can be accompanied by the emission of a characteristic X-ray (Fig. 3.7a) or by the ejection of an second electron to which the energy is non-radiatively transfered from the filling electron (Fig. 3.7b) [91, 99]. The latter process is called Auger effect with the ejected electron having a characteristic kinetic energy defined by the energy levels involved and independent of the enegy $h\nu$ of the incident light. If the hole is filled in the K-shell from an electron of the L-shell as shown in Fig. 3.7b and an electron from the L-shell is ejected, then the latter is called a KLL electron. Its energy is equal to the recombination energy but reduced by the sum of the work function and its binding energy. In the present work, Auger electron spectroscopy has been used qualitatively to check for the cleanliness of the prepared surfaces and also quantitatively to analyze relative surface coverages. This can be done using the peak to peak height of a differentiated Auger peak, which is proportional to the number of excited atoms [100].

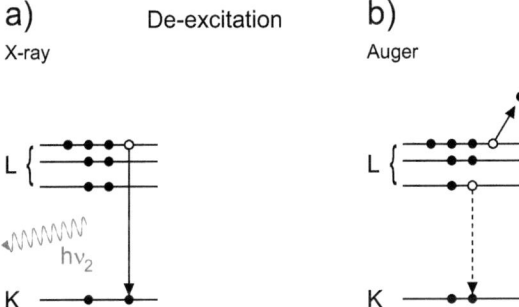

Figure 3.7.: Competing de-excitation processes occurring after photoionization. a) X-ray fluorescence and b) Auger process.

3.1.6. Quantitative analysis of core level spectra

Quantitative evaluation of the XPS core level spectra is founded on the fact, that the cross section of a certain element is independent of its electronic configuration and thus constant for a given beam energy and experimental geometry [91]. Consequently, the integrated peak area A_x of a core level X is proportional to the number of atoms in the analyzed volume. In general, the integrated peak area of a element in a compound is a function of the cross section,

Chapter 3. Experimental methods

the detection efficiency and transmission function of the spectrometer, the incident photon flux, and n_x, which is the density of the elements in a certain volume [91]. However, for a practical analysis all factors except for n_x are summarized in a so called atomic sensitivity factor (ASF):

$$A_x = n_x \cdot ASF_x \qquad (3.20)$$

and

$$n_x = \frac{A_x}{ASF_x} \qquad (3.21)$$

The ASFs should be ideally determined from a sample with well known stochiometric ratio of the elements using the experimental setup, which is also being used for the experiments. However, this is not always possible and thus the ASF values for special geometries, certain types of analyzers and light sources are tabulated for a large number of elements[6] [101]. The percentage C_x of a specific element in a homogeneous sample is given by (Eq. 3.21):

$$C_x = \frac{n_x}{\sum_i n_i} = \frac{\frac{A_x}{ASF_x}}{\sum_i \frac{A_i}{ASF_i}} \qquad (3.22)$$

3.2. Infrared spectroscopy

The analysis of the interaction of light in the infrared range with matter is called infrared spectroscopy. To obtain information on the chemical structure of molecules, the absorption as a function of the wavenumber[7] of the incident infrared light by (in most cases) organic materials is studied. Chemical bonds vibrate at specific frequencies, which are very sensitive to the structure and conformation of the molecules. The vibrations correspond to distinct energy levels, which can be excited by the right (resonant) infrared (IR) radiation.

3.2.1. Molecular vibrations

For diatomic molecules in the gas phase only one vibrational motion of the constituting atoms is possible, which is bond stretching. The potential $V(r)$ between the atoms of the diatomic molecule with the distance r can be modeled by the anharmonic Morse-potential [102]. It has been found empirically and accounts for the fact, that a certain distance the molecule will dissociate. In the following, only the harmonic approximation shall be discussed, which

---

[6]Their use has shown a reproducibility in quantification for well known components of about 10% [91].
[7]The wavenumber is the reciprocal of the wavelength and is generally expressed in cm^{-1}.

3.2. Infrared spectroscopy

is a good description for the Morse-potential only very close to the minimum. The potential in the harmonic approximation is given by [103]:

$$V(r) = k_s/2(r - r_0)^2 \tag{3.23}$$

where r_0 is the equilibrium distance and k_s the spring constant of the bond. From the quantum-mechanical solution of the Schrödinger-equation, the vibrational frequency ν can be obtained [103]:

$$\nu = \tilde{\nu} \cdot c = \frac{1}{2\pi}\sqrt{\frac{k_s}{m_{red}}} \tag{3.24}$$

where $\tilde{\nu}$ denotes the wavenumber of the vibration, c the speed of light and m_{red} the reduced mass of the two atoms. Furthermore, it can be shown that the energy of the vibration $E(v)$ can only take discrete values:

$$E(v) = \tilde{\nu} \cdot hc = \frac{h}{2\pi}\nu\left(v + \frac{1}{2}\right), \quad v = 0, 1, 2, 3... \tag{3.25}$$

with v being the vibrational quantum number and h Planck's constant. This shows that in the harmonic approximation all vibrational states are equidistant, which is not the case for the Morse-potential. There, the energetic spacing between two states v and $v\prime$ becomes less with increasing v.

IR light absorption only occurs, when the energy of the light matches the energetic spacing between two adjacent states v and $v\prime$ of the harmonic oscillator. Thus, the vibrational quantum number v can only change by [93, 102]:

$$\Delta v = \pm 1 \tag{3.26}$$

This is one of the *two* selection rules in infrared spectroscopy. It is slightly modified in the case of the Morse-potential, where also higher overtones ($\Delta v = \pm 2, 3, ...$) are excitable. Their intensity relative to the fundamental transition ($\Delta v = \pm 1$) is, however, rather small and thus overtones only appear (if at all) with a low intensity in the spectra. For example, the ratio between fundamental and first overtone transition for the stretching vibration of HCl (2900 cm^{-1}) is 1:10^{-3} [102].

A molecule consisting of N atoms generally has $3N$ degrees of freedom, because every single atom of the molecule contributes three translational degrees of freedom. For the whole molecule these $3N$ are partitioned on three translational degrees (x, y, z) and another three on the rotation around the translational axis leaving $3N - 6$ degrees of freedom for vibrational modes. Linear or diatomic molecules have only two rotational degrees of freedom as the rotation around the linear axis leaves the molecule unchanged. If the molecules are adsorbed

Chapter 3. Experimental methods

on surfaces, they lose their translational and rotational degrees of freedom and thus $3N$ vibrational modes are available. The translational and rotational degrees of freedom are converted into vibrational degrees of freedom, which are then generally called frustrated translations and rotations. The complex motion of atoms in a molecule is usually described using normal coordinates, which are chosen such that the different vibrational modes are decoupled [102]. Note that for a molecule in the gas-phase, this also eliminates translations and rotations, as all motions are described with respect to the center of mass. These decoupled vibrations are called *normal modes* and can be excited independently. In general, all atoms of a molecule are involved in each normal mode. However sometimes some atoms remain fixed. Furthermore, the normal modes leave the center of mass of the molecule unchanged. The normal modes of a molecule can be divided into four classes of vibrational motion [104] as visualized in Fig. 3.8:

- Stretching, where the vibrating bonds are elongated along the direct connection between two atoms. This vibration takes place in the plane (ip) between the two atoms. If more than two atoms are contributing, the vibration can be further divided into symmetric and antisymmetric stretchings.

- Bending, where the vibration changes the angle between (three) atoms. This vibration also takes places in the plane containing the atoms.

- Torsion, where the vibration changes the dihedral angle between two planes containing four atoms. This is an out of plane (oop) vibrational motion.

- Wagging, which is an out of plane bending vibration of three or more atoms.

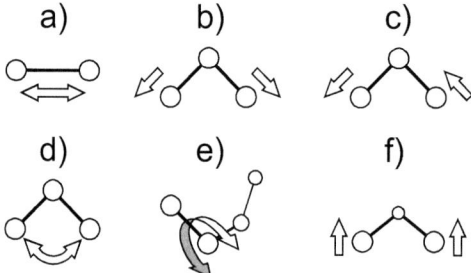

Figure 3.8.: Classes of vibrational motion. a) bond stretching, b) symmetric and c) antisymmetric stretching of a atomic group such as CH_2, d) bending, e) torsion, and f) wagging vibrations.

3.2. Infrared spectroscopy

The excitation of any of these modes by IR light is governed by selection rules for electronic dipole transitions. One was already given, stating that only changes in the vibrational quantum number by unity are allowed in the harmonic approximation. The second requirement for a transition to be IR active, is that the incident IR light can couple to the initial and final state of the transition [92]. This transition probability $P_{v \to v\prime}$ of a pure vibrational transition v is proportional to the square of the transition dipole moment $M_{v \to v\prime}$, which is given by:

$$M_{v \to v\prime} = |\langle \Psi(v) | \underline{\mu} | \Psi(v\prime) \rangle|^2 \tag{3.27}$$

where $\Psi(v)$ and $\Psi(v\prime)$ are the vibrational initial state with quantum number v and final state with quantum number $v\prime$ and $\underline{\mu} = \vec{A_0} \cdot \underline{P}$ is the dipole moment operator of the incident light. $\underline{\mu}$ can be split into the three spatial components $\underline{\mu_x}$, $\underline{\mu_y}$, and $\underline{\mu_z}$ and thus Eq. 3.27 can be written for each component:

$$M^x_{v \to v\prime} = |\langle \Psi(v) \underline{\mu_x} \Psi(v\prime) \rangle|^2 \tag{3.28}$$

$$M^y_{v \to v\prime} = |\langle \Psi(v) | \underline{\mu_y} | \Psi(v\prime) \rangle|^2 \tag{3.29}$$

$$M^z_{v \to v\prime} = |\langle \Psi(v) | \underline{\mu_z} | \Psi(v\prime) \rangle|^2 \tag{3.30}$$

For a transition to be dipole active, at least one of these integrals must be nonzero resulting in Eq. 3.27 to be nonzero. Conditions for one of the component integrals to be nonzero can be found by analyzing the product in the integrand $\Psi(v)\underline{\mu_x}\Psi(v\prime)$ (the same applies for y and z). Using group theory it is found that this product needs to be totally symmetric (irreducible representation: usually A^8) [93, 92]. In general one excites from the ground state $\Psi(v)$ (which is totally symmetric [92]) because the energetic spacing of the vibrational levels is sufficiently large compared with the thermal energy $k_B T$ at room temperature, so that only the ground state will be significantly populated. Consequently, the integrand is totally symmetric when the product of $\underline{\mu_x}$ and $\Psi(v\prime)$ is totally symmetric [92, 105]. The components ($\underline{\mu_x}$, $\underline{\mu_y}$, and $\underline{\mu_z}$) of the dipole moment operator transform exactly as the corresponding coordinates in space (x,y,z). Therefore, they have the same irreducible symmetry representations and can be looked up for each point group in the character tables listed in the appendix A.1. By forming the direct product of the two irreducible representations for $\underline{\mu_x}$ and $\Psi(v\prime)$ it can thus be concluded which integral of the transition dipole moments $M^x_{v \to v\prime}$, $M^y_{v \to v\prime}$, and $M^z_{v \to v\prime}$

---

[8]The sets of characters for all symmetry operations of a point group are referred to as irreducible representations of the particular group [92]. They indicate how a molecular property (translation, operators, etc.) transforms under all symmetry operations of the point group. For the cases used in this work, they are listed in the character tables in the appendix A.1

Chapter 3. Experimental methods

is nonzero, just by evaluation of the symmetry of the integrand (further details see Refs. [92, 105]). It should be noted that the excitation of a molecular vibration also depends on the polarization of the incident light and the orientation of the corresponding transition dipole moment. If the projection of the polarization vector onto the vector of the transition dipole moment is zero, then this mode will not be seen in the spectrum.

In the case of molecules adsorbed on metallic substrates an additional selection rule applies, which is the so called "surface selection rule". It further restricts the excitable vibrational modes to only those, who have a non vanishing transition dipole moment $M_{vv'}^z$ in the z-direction (normal to the surface). Further explanations will be given in detail in the next section.

The intensity I of an allowed transition is proportional to the square of the derivative of the dipole moment $\underline{\mu}$ with respect to the coordinates x, y, and z [106]:

$$I_k \propto \left(\frac{\partial \mu}{\partial k}\right)^2 \qquad k = x, y, z \tag{3.31}$$

In the absence of intermolecular interactions, the Intensity I of a certain mode should linearly increase with the number of molecular dipoles in the beam [105].

3.2.2. Vibrations at surfaces

Absorption of IR radiation by molecules in the gas phase depends, as stated earlier, entirely on the polarization of the incident light and the orientation of the transition dipole moment of the molecules. Molecules adsorbed on clean metal surfaces can be probed by IR radiation in a specular reflection geometry as shown in Fig. 3.9. Here, the absorption of light by the molecules is governed by the dielectric behavior of the metal because both the electric field of the incoming light and the transition dipole moment of the molecules, will interact with the metal electrons [106]. The intensity of an absorption band in the spectrum depends on the strength of the electric field at the surface, the orientation of the molecular transition dipole moment and the number of molecules probed by the IR beam. Information about the binding of the molecules to the surface and their orientation with respect to the surface normal can be obtained as a result of the strict *surface selection rule* applicable to the reflection of IR radiation at metallic substrates. In the first part of this subsection, it shall be shown why grazing incident angles are necessary on clean metal surfaces and illustrated how the *surface selection rule* is obtained. In the second part, it will be demonstrated that this still holds if the metal surface is covered by an adsorbate using a three-phase model.

3.2. Infrared spectroscopy

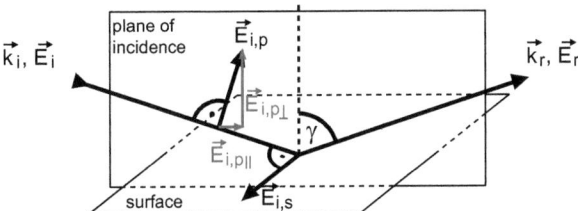

Figure 3.9.: Optical geometry in a RAIRS experiment. The electric field of the incident light ($\vec{E_i}$) can be separated into a component polarized parallel $\vec{E_{i,p}}$ and one perpendicular $\vec{E_{i,s}}$ to the plane of incidence. The parallel component is further separated into a component parallel $\vec{E_{i,p\parallel}}$ and one perpendicular $\vec{E_{i,p\perp}}$ to the surface. The angle between incident/reflected light and the surface normal is γ.

Reflection at the clean metal surface

As shown in Fig. 3.9, the electric field of the incident light ($\vec{E_i}$) can be separated into a component $\vec{E_{i,p}}$ polarized parallel to the plane of incidence and one, $\vec{E_{i,s}}$ that is polarized perpendicular. The reflected intensities R_p and R_s and the phase shifts of the reflected electric fields, δ_p and δ_s, of both components are defined by Fresnel's Equations [107, 108, 109, 110]. If one uses the complex refractive index $\tilde{n} = n + ik$ and the assumption that $n^2 + k^2 \gg 1$, which is valid for radiation in the infrared range [107], the reflected intensities for light polarized parallel R_p and perpendicular R_s to the plane of incidence are:

$$R_p = r_p^2 = \left|\frac{\vec{E_{r,p}}}{\vec{E_i}}\right|^2 = \frac{(n - \cos\gamma)^2 + k^2}{(n + \cos\gamma)^2 + k^2} \qquad (3.32)$$

$$R_s = r_s^2 = \left|\frac{\vec{E_{r,s}}}{\vec{E_i}}\right|^2 = \frac{(n - \sec\gamma)^2 + k^2}{(n + \sec\gamma)^2 + k^2} \qquad (3.33)$$

The phase shifts are then given by:

$$\delta_p = \arctan\left(\frac{\mathrm{Im}(r_p)}{\mathrm{Re}(r_p)}\right) \qquad (3.34)$$

$$\delta_s = \arctan\left(\frac{\mathrm{Im}(r_s)}{\mathrm{Re}(r_s)}\right) \qquad (3.35)$$

Accordingly, the phase shift of light polarized perpendicular to the plane of incidence is approximately 180° for all γ as plotted in Fig. 3.10 [111]. As the reflection coefficients r_p and r_s must obviously be the same for $\gamma = 0°$ (normal incidence), also the phase shifts δ_p and δ_s

Chapter 3. Experimental methods

must be the same. Note that most of the reports in literature show the wrong behavior here [105, 106, 110]. In contrast to δ_s, δ_p has a strong dependence on γ and changes by almost 180° going from normal incidence to $\gamma = 90°$.

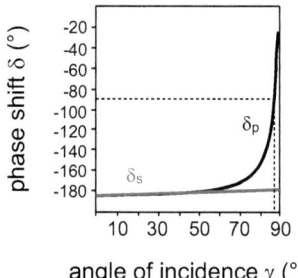

Figure 3.10.: Phase shift δ of the electric field components parallel (δ_p) and perpendicular (δ_s) plotted versus the incident angle γ upon reflection at a metal surface.

If the amplitude of an incoming electric field $\vec{E}_i = \vec{E}_0 e^{i\varphi}$ with an arbitrary phase φ is considered, then the amplitude of the reflected field is given by $\vec{E}_r = r \cdot \vec{E}_0 e^{(i\varphi + \delta)}$. The resulting field $\vec{E}$ at the surface is then the superposition of both incoming and reflected fields:

$$\vec{E} = \vec{E}_i + \vec{E}_r = \vec{E}_0 (e^{i\varphi} + r \cdot e^{i\varphi + \delta}) \tag{3.36}$$

This holds for both electric field components, the one polarized parallel ($\vec{E} = \vec{E}_p$) and the other polarized perpendicular ($\vec{E} = \vec{E}_s$) to the surface. The fields of the latter ($\vec{E}_{i,s}$ and $\vec{E}_{r,s}$) are parallel to the surface for all angles of incidence γ. Thus the resulting electric field perpendicular to the plane of incidence $\vec{E}_s$ can be written as:

$$\vec{E}_s = \vec{E}_{i,s}(e^{i\varphi} + r_s \cdot e^{i\varphi + \delta_s}) \tag{3.37}$$

For metals like gold, silver, and copper the reflectivity r in the infrared range is close to unity [107, 112] and thus $r_p = r_s \approx 1$. For a phase shift $\delta_s \approx -180°$ ($-\pi$) the electric field $\vec{E}_s$ at the surface almost vanishes (for all angles γ).

The phase shift δ_p strongly depends on the angle of incidence. For δ_p around 90°, which is reached at incident angles γ of about 87° (as indicated by the dashed lines in Fig. 3.10), both amplitudes $A_{i,p}$ and $A_{r,p}$ add constructively to give an elliptical standing wave at the surface ($\vec{E}_p$) with a large component normal to the surface. This is depicted in Fig. 3.11, where the sum of both incident and reflected amplitudes is shown at different times (assuming they have a similar magnitude). Note that the component parallel to the surface is largely over-sized

3.2. Infrared spectroscopy

in Fig. 3.11.

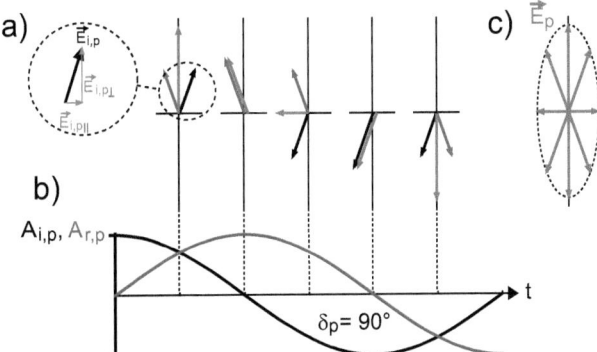

Figure 3.11.: Generation of an elliptical standing electric wave $\vec{E}_p$ parallel to the surface. a) vector-addition of the incident electric field component parallel to the plane of incidence $\vec{E}_{i,p}$ and the reflected electric field component parallel to the plane of incidence $\vec{E}_{r,p}$. The result is the electric field vector of the elliptical standing electric wave $\vec{E}_p$ at the surface. The addition is shown at different times. b) amplitudes of the electric fields $\vec{E}_{i,p}$ and $\vec{E}_{r,p}$ ($A_{i,p}$ and $A_{r,p}$) plotted versus time for a phase shift of $\delta_p = 90°$. c) resulting elliptical standing wave $\vec{E}_p$. Note that the component $\vec{E}_{i,p\perp}$ is largely over-sized in a) and c).

The added electric fields at the surface for the two components of $\vec{E}_{i,p}$ and $\vec{E}_{r,p}$ in the plane of incidence can be written as:

$$\vec{E}_{p\parallel} = \vec{E}_{i,p\parallel}(e^{i\varphi} + r_p \cdot e^{i\varphi+\delta_p}) \tag{3.38}$$

$$\vec{E}_{p\perp} = \vec{E}_{i,p\perp}(e^{i\varphi} + r_p \cdot e^{i\varphi+\delta_p}) \tag{3.39}$$

The amplitudes A_s, $A_{p\parallel}$, and $A_{p,\perp}$ of the three electric field components present at the surface ($\vec{E}_s$, $\vec{E}_{p\parallel}$, and $\vec{E}_{p\perp}$) expressed in terms of A_i have been evaluated in detail by Hollins and Pritchard [110]. They are plotted for $n = 3.5$ and $k = 30$ in Fig. 3.12. The figure shows that the amplitudes of the electric fields $\vec{E}_s$ and $\vec{E}_{p\parallel}$ are negligible for all angles of incidence γ and that $\vec{E}_{p\perp}$ strongly peaks around $\gamma = 75° - 80°$ (Fig. 3.12a). The specular geometry of the RAIRS experiment leads to an increase of the probed area (and thus the number of probed molecules) scaling with the secans (sec) of the incident angle γ. The intensity of the molecular absorption is proportional to the square of the amplitudes of the electric fields at the surface. Putting both together, a surface intensity function $(A_k/A_i)^2 \cdot \sec(\gamma)$ with $k = s, p \parallel$, and $p \perp$

43

Chapter 3. Experimental methods

is obtained. This function (as shown in Fig. 3.12b) strongly peaks close to grazing incidence for $\vec{E}_{p\|}$ and $\vec{E}_{p\perp}$. However, the intensities of the former ($\vec{E}_{p\perp}$) and $\vec{E}_s$ are by a factor of 1000 smaller compared to $\vec{E}_{p\perp}$. The different maxima in Figs. 3.12a and b show that there is generally a small trade-off between maximized electric field and intensity at the surface. In this work all experiments are performed at an angle $\gamma = 83°$.

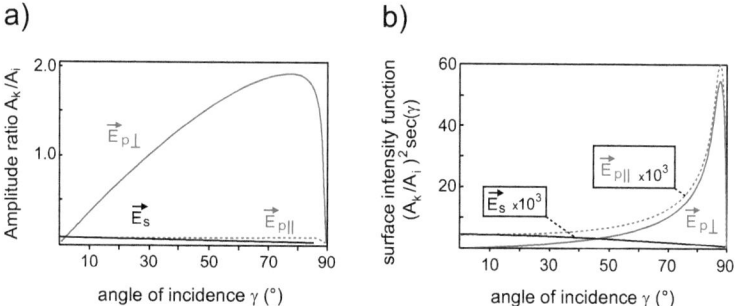

Figure 3.12.: a) amplitude ratios and b) surface intensity function of the surface electric fields $\vec{E}_s$, $\vec{E}_{p\|}$, and $\vec{E}_{p\perp}$ as a function of γ. They are plotted for $n = 3.5$ and $k = 30$ in according to Hollins and Pritchard [110].

The above considerations for the clean metal surface show that mainly an electric field with a component perpendicular to the surface will be established at the surface. From this stems the *surface selection rule* in RAIRS. Only vibrations which have their transitions dipole moment perpendicular to the surface (in z-direction) can be excited.

Reflection at the adsorbate covered surface

In the experiment, the surface will be covered by an (organic) adsorbate, which further complicates the theoretical description. The reflectance of the clean metal surface, R_0, is generally taken as a reference. When the surface is covered by the adsorbate, the reflectance will change by $\Delta R = R_0 - R_A$, where R_A is the reflectance of the adsorbate covered surface as shown in Fig. 3.13a. The theoretical description of the reflectance change has been treated by Greenler, using the Fresnel equations for a three-phase model, as shown in Fig. 3.13b [108]. For each phase a complex isotropic dielectric constant $\epsilon = (n + ik)^2$ is introduced, which is certainly an approximation as sharp interfaces will probably not exist, especially between the organic film and the vacuum.

The basic Fresnel equations for the reflectance have been greatly simplified by Aspnes and McIntyre using a linear approximation when expanding in powers of $d/\lambda \ll 1$ for film

3.2. Infrared spectroscopy

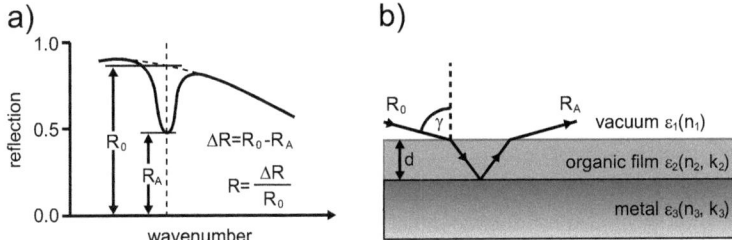

Figure 3.13.: a) definition of the reflectance change occurring for the reflection of IR light at an adsorbate covered surface. R_0 is the reference of the uncovered and R_A the reflectance of the adsorbate covered surface. The reflectance change is defined as $\Delta R = R_0 - R_A$. b) schematic drawing of the adsorbate (layer thickness d) covered surface used for the three phase model. Each phase (metal, organic film, and vacuum) is characterized by a complex isotropic dielectric constant $\epsilon = (n + ik)^2$ and sharp interfaces are assumed.

thicknesses of $d = 100 - 1000$ Å [106, 109]. Then the reflectance for p- and s-polarized light are given by:

$$R_p = \frac{\Delta R_p}{R_{p0}} = \frac{8\pi n_1 d}{\lambda} \cos\gamma \operatorname{Im}\left(\frac{\epsilon_2 - \epsilon_3}{\epsilon_1 - \epsilon_3}\right) \left[\frac{1 - \frac{\epsilon_1}{\epsilon_2\epsilon_3}\epsilon_2\epsilon_3(\sin\gamma)^2}{1 - \frac{1}{\epsilon_3}(\epsilon_1 + \epsilon_3)(\sin\gamma)^2}\right] \quad (3.40)$$

$$R_s = \frac{\Delta R_s}{R_{s0}} = \frac{8\pi n_1 d}{\lambda} \cos\gamma \operatorname{Im}\left(\frac{\epsilon_2 - \epsilon_3}{\epsilon_1 - \epsilon_3}\right) \quad (3.41)$$

Since for the first phase (vacuum) the dielectric constant reduces to $\epsilon_1 = n_1^2 = 1$, the above equations can be further simplified. Following Hoffmann [106] who states that for IR light $\epsilon_3 \gg \epsilon_2 \approx 1$ for organic adsorbates and $d/\lambda \ll 1$, one sees immediately that the reflectance change for s-polarized light becomes negligibly small. This is a further evidence for the strict validity of the surface selection rule. According to Ibach [113] the expression for p-polarized light can be further simplified because $\epsilon_3 \gg \epsilon_2$ and $(\cos\gamma)^2 \gg \frac{1}{\epsilon_3}$ generally holds for most experiments. Combining all approximations, Eq. 3.40 simplifies to:

$$R_p = \frac{\Delta R_p}{R_{p0}} = \frac{8\pi d(\sin\gamma)^2}{\lambda \cos\gamma} \operatorname{Im}\left(-\frac{1}{\epsilon_2}\right) \quad (3.42)$$

Similar to the results obtained for reflection at the clean metal surface (Eq. 3.39), Eq. 3.42 implies that experiments have to be performed at high incident angles. Furthermore, the results of this classical treatment of the reflectivity at the surface again clearly imposes a strict surface selection rule on the vibrations that are excitable in RAIRS experiments. It is, however, also possible to arrive at this surface selection rule using image dipole theory

Chapter 3. Experimental methods

as explained in appendix A.2. Using the surface selection rule it can be concluded that the relevant transition dipole moment for vibrations on surfaces is that of the z-component: $M^z_{vv'}$ (if the molecule is lying flat). Only those vibrations will be observed in the spectra, for which the product of $\underline{\mu}$ and $\Phi(v')$ in the integrand of the component's transition dipole moment yields an irreducible representation that transforms as z [105] (taking again the initial state as the ground state). This does not necessarily mean that the corresponding vibration (and the collective motion of the atoms) is directed normal to the surface. This is because the transformation properties (the irreducible representation) of the dipole moment operator $\underline{\mu}_z$ are important for the non vanishing transition dipole moment and not the atomic displacements. As long as the vibration belongs to the totally symmetric representation, it is allowed by the selection rules and will thus be observed in the spectra, even if the vibrational motion is parallel to the surface. This can be understood by considering a vibration *parallel* to the surface, which alters the amount of charge transferred *normal* to the surface and thus creates a dynamical dipole moment normal to the surface[113]. However, the intensity of an IR band in the spectrum is related to the strength of this dynamic dipole, which group theory does not account for. The observation of molecular vibrations, formally allowed by the selection rules therefore critically depends on the magnitude of the perpendicular component of the dynamic dipole. The latter can be rather weak in some cases and consequently not all allowed transitions may be observed in the spectra [113, 114].

Further complication of the interpretation of RAIRS spectra stems from the fact, that the symmetry of the adsorption site of the molecule on the surface needs to be taken into account in the case of strong interaction (chemisorption). This will in general reduce the symmetry of the point group and change the number of allowed vibrational modes[9]. Using correlation tables it is possible to determine how the representations of a certain mode change when the symmetry of the point group is changed. How they can be obtained is described in detail in Ref. [113] and the cases needed in this work are reproduced in Sec. A.1 of the appendix. If the interaction of the molecule with the substrate is rather weak (physisorption), the symmetry of the gas-phase point group is retained, which is in general also the case for the bulk film.

3.2.3. Information content of vibrational modes of adsorbed molecules

The vibrational frequency, intensity, and line shape of a mode of an adsorbed species are generally changed compared to their gas phase analogous. From this characteristic change, a wealth of information can be extracted if the involved processes can be separated and the molecule is not too complex. The investigated systems in this work, however, are rather complex and thus the analysis of the different processes occurring on the substrate cannot unequivocally be resolved. Nevertheless, a brief overview of the different processes leading to

---
[9]For example the degeneracy of gas-phase modes could be lifted.

3.2. Infrared spectroscopy

the mentioned changes shall be given here.

Vibrational frequency shifts

The vibrational frequency of a single adsorbed species is usually shifted compared to its gas phase value. One of the reasons is mechanical renormalization resulting in a shift of the gas phase frequency to higher wavenumbers, because the vibrating bond is now attached to a rigid substrate, which can be considered as an additional spring [106]. Moreover, the interaction of the vibrating dipole with its own image in the metal can reduce the frequency (compared to the gas phase value) as theoretically described by Efrima and Metiu in Refs. [115, 116]. Chemical shifts of a vibrational mode occur as a result of the bonding of a molecule to the substrate and the accompanying charge transfer. This charge transfer results in a donation or acceptance of electrons by the molecule and thus changes the occupation of certain molecular orbitals as described by the Blyholder model [117]. Depending on the character of the molecular orbital - bonding or antibonding - the charge transfer will weaken or strengthen the bond. Weakened bonds will generally reduce the vibrational frequency, while strengthened bonds will increase it [106, 118]. Certainly, all of the above described processes can occur simultaneously and partially cancel each other in such a way, that the individual contributions can no longer be separated. These shifts occur even in the absence of molecule-molecule interactions.

If the coverage is increased and the molecules on the substrate become denser, their interaction can lead to further shifts of the vibrational frequency. Two main contributions are distinguished in literature [105, 106]: dipole coupling of two adjacent molecules and static shifts, which comprises the coverage dependency of the chemical shift and the static dipole contributions. In the first case the electric field of a molecular dipole is modified by the presence of the neighboring molecular dipoles. The extend of this dipole coupling depends on the magnitude of the polarizability of the molecules and their density on the surface. Usually this results in an upward frequency shift with increasing coverage. The origin of a chemical shift has been described in the last paragraph as the charge transfer between molecule and metal. The coverage dependency arises from the fact, that the amount of charge transfered usually changes when the density of molecules is increased, since a competition for the metal electrons among the molecules occurs [105, 117]. As discussed before, depending on the direction of the charge transfer, the frequency is either shifted up- or downwards. Another factor for a frequency shift at higher coverages is intermolecular repulsion. With increasing coverage the wave functions of adjacent molecules overlap and as a result their occupation might change [106, 114, 119].

Chapter 3. Experimental methods

Intensity changes

The intensity I of an infrared absorption in the absence of intermolecular interactions, as stated before, should linearly increase with increasing amount of molecules on the surface. However when intermolecular interactions are present depolarization can take place, which reduces the integrated absorption intensity I of a certain vibration for higher coverages [106].

Line shapes and half-width

The change in line shape and half-width of a vibrational mode of an adsorbed molecule compared to its gas phase value can be ascribed to homogeneous and inhomogeneous line broadening mechanisms [105, 106]. Homogeneous line broadening is generally caused by the damping of the vibrational mode, which reduces the lifetime of the excited vibrational state. The damping needs a process of dissipative energy decay, which can occur by phonon coupling or through the creation of an electron-hole pair. However, also other pathways are possible as discussed in detail in Ref. [105]. Inhomogeneous line broadening results from the inhomogeneous distribution of individual oscillator frequencies. This is determined either by different molecular adsorption sites or different intermolecular interactions due to disorder in the molecular layer and consequently different intermolecular distances. In the case of perfect order, inhomogeneous broadening should be negligible. An overview of different line shapes is given in Ref. [106].

3.2.4. Fourier-transform infrared spectroscopy

Different techniques exist for measuring the absorption of infrared light. In dispersive spectrometers, the absorption is measured independently for every wavenumber using a grating and slit as monochromator. In contrast, Fourier-transform infrared (FTIR) spectrometers measure all wavelengths at once, which significantly reduces the time to obtain one full spectrum. Consequently, more scans can be co-added and then averaged in a certain time, which reduces the signal to noise ratio by $\sqrt{n}$ (with n as the number of scans). Other advantages compared to dispersive techniques are the constant light intensity and spectral resolution for the whole spectrum. In FTIR spectroscopy, the central part of the spectrometer is the Michelson interferometer. Here the beam coming from the light source is split in two by a semi-transparent mirror (beam splitter). One of the beams is reflected at a fixed mirror and the other one at a movable mirror as shown in Fig. 3.14.

After being reflected, both beams will interfere depending on the optical path difference δ between both interferometer arms. For a monochromatic light source with the frequency ν a cosine modulation of the intensity $I(\delta)$ as function of δ is obtained at the detector (called

3.2. Infrared spectroscopy

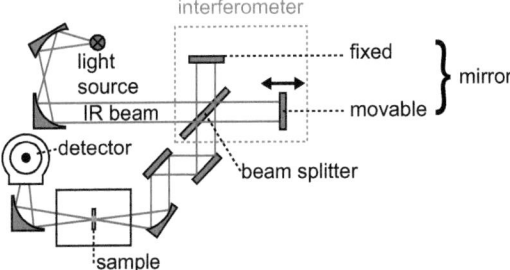

Figure 3.14.: Schematic drawing of the interferometer used in a Fourier transform (IR) spectrometer consisting of a beamsplitter, a moveable and a fixed mirror. For illustration all other parts of the spectrometer are also indicated.

interferogram) according to:

$$I(\delta) = \frac{1}{2}I(\nu)\cos(2\pi\nu\delta) \tag{3.43}$$

When the light source emits a continuous spectrum, the recorded interferogram is represented as an integral over the full spectral range:

$$I(\delta) = \int_{-\infty}^{\infty} I(\nu)\cos(2\pi\nu\delta)\,d\nu \tag{3.44}$$

The corresponding function $I(\nu)$ can be obtained by cosine Fourier transformation, which yields the full range spectrum:

$$I(\nu) = \int_{-\infty}^{\infty} I(\delta)\cos(2\pi\nu\delta)\,d\delta \tag{3.45}$$

It should be noted, that $I(\delta)$ is an even function so that we may also write 3.45 as:

$$I(\nu) = 2\int_{0}^{\infty} I(\delta)\cos(2\pi\nu\delta)\,d\delta \tag{3.46}$$

Using Eq. 3.45, the complete spectrum could be obtained at infinitely high resolution. However, this would imply that δ could be varied from 0 to ∞, which is practically not possible. Consequently, a limited resolution is enforced by scanning the moving mirror only over a finite distance and thus varying δ only over a finite range. Therefore, different apodization functions are generally used in FTIR spectroscopy to make up for this, where the most prominent ones are the triangular and the Happ-Genzel apodization functions (the latter has been

49

Chapter 3. Experimental methods

used for all IR spectra shown in this work). A detailed discussion of the different apodization functions can be found in Ref. [111].

Chapter 4.
Materials and experimental setups

In this chapter, the organic materials and the experimental setups used in this work will be described. In addition, experimental parameters and details of the data evaluation will be illustrated. The chapter will conclude with a brief introduction of the theoretical methods applied within this work.

4.1. Materials

This section is divided into two parts. First the molecular electron donor and acceptor materials will be briefly introduced. In the second part the hole and electron transport materials will be described, that have been used to quantify the changes in the electronic level alignment due to the insertion of donor or acceptor layers.

4.1.1. Molecular electron donor and acceptor materials

To change the energy level alignment of subsequently deposited organic materials, strong electron donor and acceptor materials are necessary. These molecules should undergo a strong charge transfer with the substrates to increase the work function (acceptors) or decrease the work function beyond the "push-back" effect (donors) relative to the value of the pristine substrate. Therefore, molecules have been chosen and synthesized, which are especially electron poor (acceptors) or electron rich (donors). Because this is in most cases not obvious from the chemical structure alone, it has been proven valuable in many cases to choose functional groups depending on their empirical Hammett parameter σ [120]. σ is a measure of a functional group to increase or decrease the electron density of adjacent aromatic systems. A compilation of values for various functional groups is given in Ref. [121]. The investigated molecules used in this work were synthesized by R. Rieger (Max Planck Institut für Polymerforschung, Mainz), except for 9,9'-ethane-1,2-diylidene-bis(N-methyl-9,10-dihydroacridine) (NMA), which was synthesized by L. Beverina (State University of Milano-Bicocca, Milano). The chemical structures of the electron acceptor and donor materials

Chapter 4. Materials and experimental setups

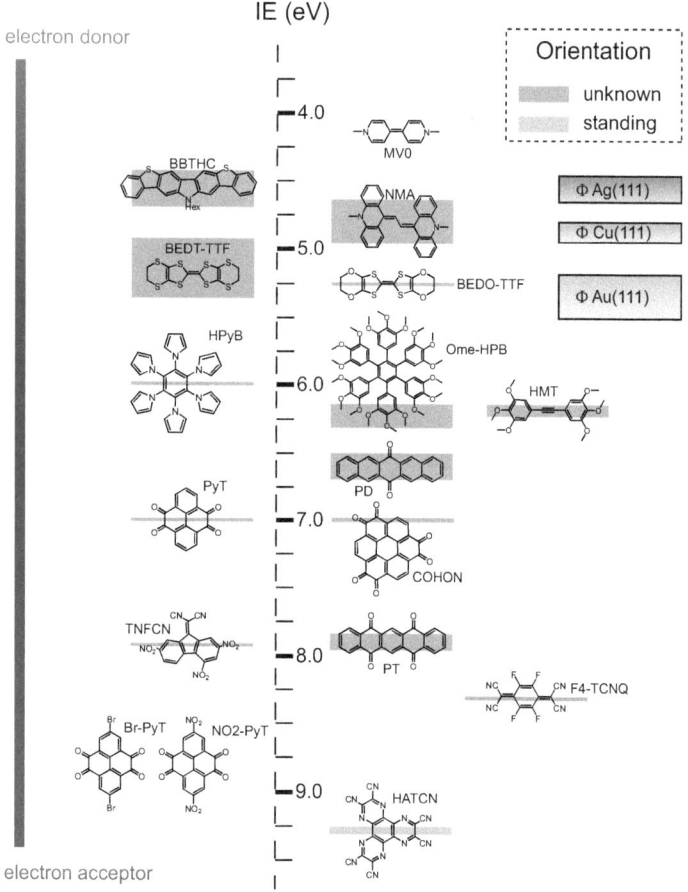

Figure 4.1.: Chemical structures of the molecules investigated in this work, sorted by their experimental IE value range. Their full names are listed in Tab. 4.1. For molecules, where the orientation with respect to the substrate surface is known, this is also indicated. For all other molecules the experimental range is indicated by the width of the bar. For MV0, BEDO-TTF, Br-PyT, and NO2-PyT no experimental IE values could be obtained, therefore they are placed by their calculated vertical IE_{vert}.

4.1. Materials

Table 4.1.: Acronym, chemical formula, name, and molecular weight of the different donor and acceptor materials initially screened in this work.

Acronym	Chemical formula	Name	Molecular weight (g mol^{-1})
MV0	$C_{12}H_{14}N_2$	1,1'-dimethyl-4,4'-bipyridinylidene	186
BBTHC	$C_{30}H_{25}N_1$	bisbenzothienylhexyl-carbazole	463
NMA	$C_{30}H_{24}N_2$	9,9'-ethane-1,2-diylidene-bis(N-methyl-9,10-dihydroacridine)	412
BEDT-TTF	$C_{10}H_8S_8$	bis(ethylene-dithiolo)tetrathiofulvalene	384
BEDO-TTF	$C_{10}H_8O_4S_4$	bis(ethylene-dioxy)tetrathiofulvalene	320
HPyB	$C_{30}H_{24}N_6$	hexapyrrolyl-benzene	468
Ome-HPB	$C_{30}H_{24}N_6$	octadecamethoxy-hexaphenylbenzene	1074
HMT	$C_{20}H_{22}O_6$	hexamethoxy-tolan	358
PD	$C_{22}H_{12}O_2$	pentacene-dion	308
PT	$C_{22}H_{10}O_4$	pentacene-tetraone	338
PyT	$C_{16}H_6O_4$	pyrene-tetraone	262
COHON	$C_{24}H_6O_6$	coronene-hexaone	390
TNFCN	$C_{16}H_5N_5O_6$	dicyanomethyl-trinitrofluorenone	263
Br-PyT	$C_{16}H_4Br_2O_4$	dibromo-pyrene-tetraone	420
NO2-PyT	$C_{16}H_4N_2O_8$	dinitro-pyrene-tetraone	352
F4-TCNQ	$C_{12}F_4N_4$	2,3,5,6-tetrafluoro-7,7,8,8-tetracyano-quinodimethane	276
HATCN	$C_{18}N_{12}$	1,4,5,8,9,12-hexaazatriphenylene-hexacarbonitrile	384

are shown in Fig. 4.1. Besides, Tab. 4.1 lists their acronym, chemical formula, name, and molecular weight.

4.1.2. Hole and electron transport materials

"Test" molecules are needed to quantify the ability of the electron accepting and donating materials to lower the hole- and electron injection barriers into subsequently deposited organic materials. In this work α–NPD has been used as a hole transport and Alq$_3$ and C$_{60}$ have been used as electron transport materials. α–NPD and Alq$_3$ exhibit an amorphous bulk structure and together with C$_{60}$ grow in a layer-by-layer fashion on various substrates at room temperature [122, 123, 124, 125, 126]. This makes them well suited as "test" molecules, because their HOMO position is not affected by an orientational dependence of the IE. The chemical structure of the "test" molecules is shown in Fig. 4.2 and the chemical formulas are listed in Tab. 4.2. They have all been obtained from Sigma-Aldrich (Taufkirchen).

Chapter 4. Materials and experimental setups

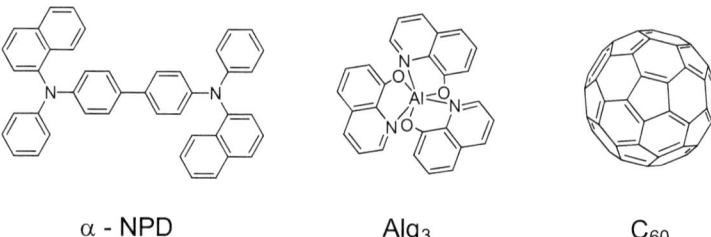

α - NPD   Alq$_3$ C$_{60}$

Figure 4.2.: Molecular structure of the hole- and electron transport materials used in this work to quantify the barrier reduction potential of the acceptor and donor materials. For full names and chemical formula see Tab. 4.2.

Table 4.2.: Acronym, chemical formula, and name of the hole- and electron transport materials used in this work.

Acronym	Chemical formula	Name
α-NPD	$C_{44}H_{32}N_2$	N,N'-diphenyl-N,N'-bis(1-naphthyl)-1,1'-biphenyl-4,4'-diamine
Alq$_3$	$C_{27}H_{18}AlN_3O_3$	8-hydroxyquinoline aluminum
C$_{60}$	C$_{60}$	Buckminster fullerene

4.2. Substrates

In this section the general cleaning protocols used for metal single crystals and indium tin oxide (ITO) substrates will be described. Reference XPS spectra as well as LEED images of the clean substrates are given.

4.2.1. Metal single crystals

In most experiments presented in this work, organic materials were evaporated onto the (111)-faces of gold, silver, and copper single crystals. They were chosen because they present a well defined surface and are also the most stable ones since they have the lowest surface energy. Furthermore, the three substrates cover a wide range of surface reactivity and work function. Both are important parameters for the molecular growth and the energy level alignment of subsequently deposited organic materials. The reactivity increases in the order gold, silver, and copper, whereas the work function of Au(111) is the highest ($\Phi_{Au} \approx 5.5$ eV), followed by Cu(111) ($\Phi_{Cu} \approx 4.9$ eV) and Ag(111) ($\Phi_{Ag} \approx 4.6$ eV). All three crystals exhibit fcc packing, which results in a hexagonal lattice for the (111)-faces. The lattice constant a of the surfaces is similar for Au(111) and Ag(111) ($a = 2.99$ Å), while it is 2.56 Å for Cu(111). All single crystals were purchased from MaTecK GmbH (Jülich). Prior to use they were cleaned by

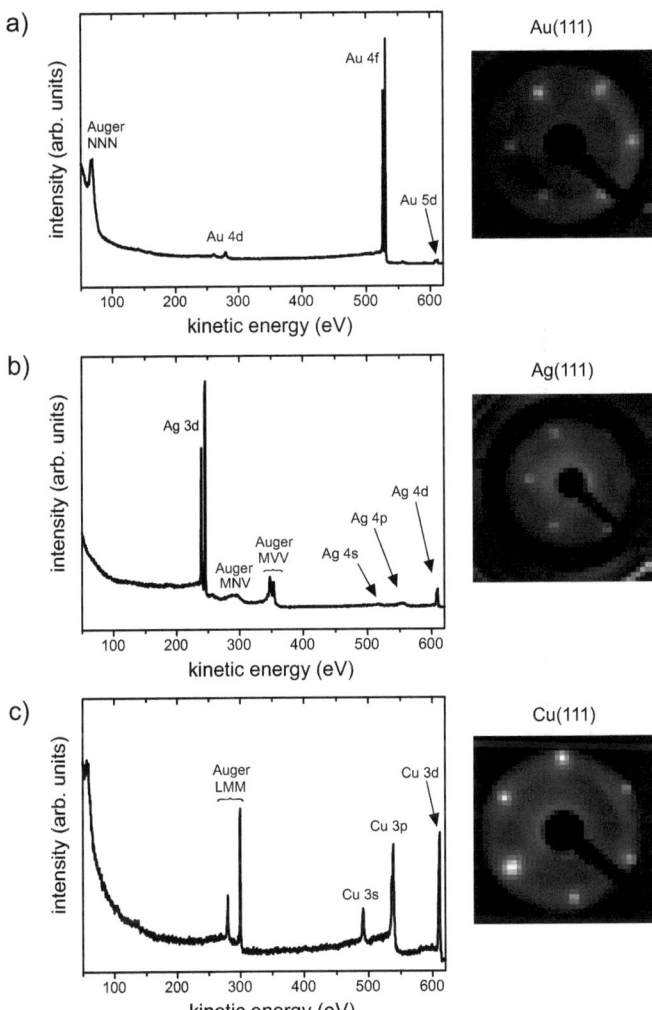

Figure 4.3.: X-ray photoelectron survey spectra ($h\nu = 620$ eV) and low energy electron diffraction images of the clean metal single crystals used in this work. The beam energies were 111.6 eV (gold), 92.4 eV (silver), and 114.1 eV (copper).

Chapter 4. Materials and experimental setups

repeated Ar-ion sputter and annealing cycles in ultra high vacuum (UHV). For sputtering in general an Ar-ion beam energy of 600 to 1200 eV was used, resulting in a current of 2 to 4 μA at the sample, which was kept for one hour. In order to reduce the roughness of the surface and to enhance surface "healing" (planarization due to atom diffusion to obtain large terraces), the single crystals were annealed after sputtering to the following temperatures: Au(111): up to 550°C, Ag(111) and Cu(111): 450°C. In addition, this was also necessary to remove implemented Ar after the sputtering. Generally several cycles of sputtering and annealing were necessary to remove all contaminants from the surface and to obtain sharp LEED spots. Cleanliness was further checked by XPS or Auger spectroscopy (depending on the experimental setup). The images and spectra of the clean surfaces are shown in Fig. 4.3. UPS spectra of the clean substrates will be given prior to each experiment.

4.2.2. Indium tin oxide

Indium tin oxide (ITO) coated glass samples were obtained from HOLST Centre (Eindhoven) and were cut into pieces of 9×9 mm^2 or 8×16 mm^2 (depending on the experimental setup). The thickness of the ITO layer was 120 nm and had a roughness of about 20 nm (rms). The sheet resistance of the ITO layer was $\approx$ 13 Ω/square. In between the ITO layer and the glass, a layer of SiO$_2$ ($\approx$ 200 nm) was incorporated as barrier. To clean the ITO substrates, they were first washed and scrubbed in an Extran solution (Merck) (6 V/V%). Afterwards they were thoroughly rinsed with copious amounts of deionized water (Milipore) and blown dry under a stream of nitrogen. In the following step the ITO substrates were immersed first in pure aceton and sonicated for 5 min. Subsequently, the ITO substrates were placed into pure isopropanol and placed into the ultrasonic bath for another 5 min. The cleaned substrates were then either used or stored in a sealed container containing pure isopropanol. ITO substrates cleaned in such a way will be refered to in this work as "solvent-cleaned". Their work function was $\approx$ 4.2 eV. Besides, in some experiments the ITO substrates were put into an UV/ozone cleaner for 30 min prior to use. They are refered to as "UV/ozone-cleaned". The work function of the UV/ozone-cleaned ITO substrates was $\approx$ 4.5 eV.

4.3. Experimental setups

4.3.1. Photoelectron spectroscopy experiments at BESSY II

The majority of the photoelectron spectroscopy data presented in this work was recorded at the synchrotron light source BESSY II (Berlin, Germany) using the endstation chamber SurICat. The electron storage ring has a circumference of 240 m and is usually operated at 1.9 GeV using an initial beam current of 300 mA. The beamline PM4 to which the chamber is connected, is located at a dipole bending magnet. As monochromator a gold coated silicon

4.3. Experimental setups

grating with a line density of 360 lines per mm was used, which allowed beam energies ranging from 18 to 2000 eV. The beam entrance slit at the beamline was chosen to be 100 µm. The spot sized at the sample position was approximately 1 mm², which is equal to the acceptance area of the hemispherical electron energy analyzer Scienta SES 100. In general two beam energies were used: 35 eV to record the valence band spectra and 620 eV to record the core level spectra. For the beamline resolution at these energies and the photon flux at the sample see Tab. A.5 in the appendix. For all spectra an analyzer pass energy of 20 eV was used, except for XPS survey spectra ($E_{pass} = 50$ eV) and for recording the SECO spectra ($E_{pass} = 5$ eV). The spectral resolution for the valence band was obtained by extrapolation of the s-state density below and the background intensity above the Fermi level as shown by the red dashed lines in Fig. 4.4. Then the energetic difference between 80% and 20% intensity drop of the metal Fermi level was taken, which yields a resolution of $\Delta E_{UPS} = 150$ meV. In the case of XPS analysis, the full width at half maximum (FWHM) of the Au 4f 7/2 line of a clean Au(111) crystal was taken as shown in Fig. 4.4. This yields a resolution of $\Delta E_{XPS} = 670$ meV.

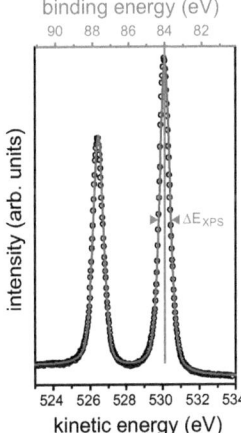
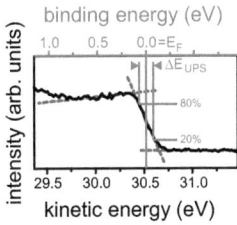

Figure 4.4.: a) XPS spectrum ($h\nu = 620$ eV) showing the Au 4f core level of a clean Au(111) single metal crystal, which has been used to convert the measured kinetic into binding energy by setting the binding energy of the Au 4f 7/2 peak to 84 eV BE. b) UPS spectra of the same metal single crystal showing the Fermi-edge, which was set to 0 eV BE in order to convert the kinetic into binding energy. In both cases the experimental resolution was obtained either from the FWHM of the Au 4f 7/2 peak or from the 80 to 20 % intensity drop at the Fermi-edge.

Chapter 4. Materials and experimental setups

The experimental setup consisted of interconnected organic evaporation (base pressure < 10^{-8} mbar), sample preparation (base pressure < 10^{-10} mbar), and analysis chambers (base pressure < 10^{-10} mbar) as shown in the schematic image in Fig. 4.5. The organic evaporation chamber was equipped with a fast-entry load lock, two positions for evaporation sources, a quarz crystal microbalance (QCM) and a sample storage system for up to four samples. In the sample preparation chamber, sputtering and annealing of the sample of up to 600 °C was possible. Furthermore, a LEED was mounted in the same chamber. The analysis chamber was equipped with the electron energy analyzer and connected to the storage ring via the beamline. The angle between analyzer and the incident beam was fixed at 60°, while the angle between the sample and the analyzer (and thus to the incident beam) was adjustable. In all spectra this angle is specified as the take-off angle α between the surface normal and the analyzer. This corresponds to an angle $\beta = 60° - \alpha$ between surface normal and the incident beam. All spectra were recorded angle-integrated with an acceptance angle of 10°. The incident light was linearly polarized in the plane of the ring. The SECO spectra were recorded with the sample biased at -10 V, in oder to clear the analyzer work function.

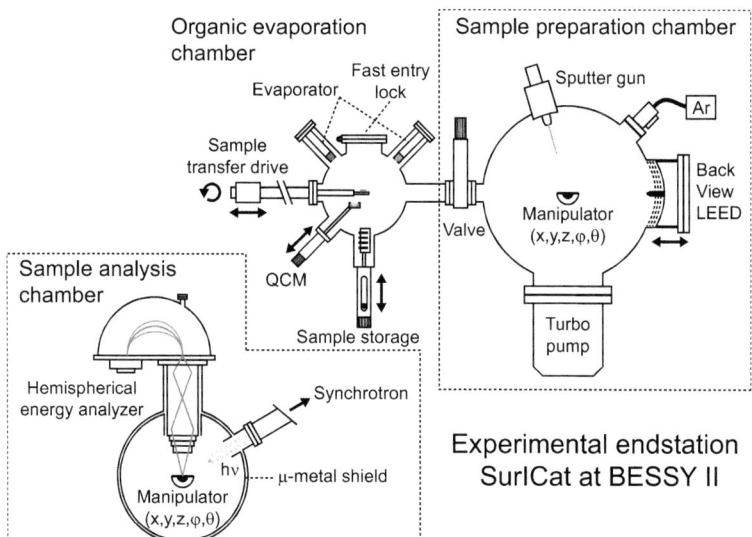

Figure 4.5.: Schematic drawing of the experimental endstation SurICat at the synchrotron light source BESSY II (top view). The sample analysis chamber is situated below the preparation chamber and the manipulator is the same in both cases. Both chambers are separated by a manual valve. QCM stands for quarz crystal microbalance and LEED for low energy electron diffraction.

4.3. Experimental setups

4.3.2. Photoelectron spectroscopy experiments at HASYLAB and Humboldt-University

Additional UPS experiments were carried out at the FLIPPER II endstation [127] (beamline E1) at the synchrotron light source HASYLAB (Hamburg, Germany) and at the Humboldt-University.

The positron storage ring DORIS III operated at HASYLAB has a circumference of 289 m, a positron energy of 4.45 GeV and an initial beam current of 120 mA. The synchrotron radiation reaching the beamline E1 is monochrimatized by a plane-grating monochromator covering the photon energy range from 17-150 eV. The endstation FLIPPER II provides two organic evaporation chambers (base pressure $< 10^{-9}$ mbar) interconnected with a sample storage, preparation and an analysis chamber (all with a base pressure $< 10^{-10}$ mbar). Furthermore, a fast entry load lock facilitates sample transfer into the UHV system, which is especially important for *ex vacuo* prepared samples such as ITO substrates. The organic evaporation chambers are each equipped with one evaporation source and a QCM. The preparation chamber provided facilities for metal single crystal cleaning (Ar-ion sputtering and annealing up to 600 °C). The analysis chamber consists of a front view LEED system and a double-pass cylindrical mirror electron energy analyzer PHI 25-260 AR (Physical Electronics), which was oriented at 90°with respect to the incident beam. For the UPS experiments, the sample surface normal was oriented at 40°with respect to the incident beam (angle between analyzer and sample surface normal = 50°). The spot size of the beam on the sample was $\approx$ 2x1 mm^2, whereas the acceptance area of the analyzer was $\approx$ 1x1 mm^2. All spectra were recorded angle-integrated with an analyzer acceptance angle of 12° - 24°and 56° - 68°. UPS spectra were recorded using a beam energy of 22 eV with the analyzer set to a constant pass energy of 15 eV. The obtained resolution measured at the Fermi-edge (as described in Sec. 4.3.1) was $\approx$ 230 meV. The photon flux at the sample is given in Tab. A.5. The SECO spectra were recorded with a sample bias of -3 to -5 V.

At the Humboldt-University a commercially available UHV system (Omicron) consisting of interconnected sample preparation (base pressure $< 10^{-9}$ mbar) and analysis (base pressure $< 10^{-10}$ mbar) chambers, was used. The preparation chamber provides facilities for metal single crystal cleaning (Ar-ion sputtering and annealing up to 600 °C), two organic evaporator sources and a QCM. The analysis chamber was equipped with a scanning tunneling/atomic force microscope (VT-STM/AFM, Omicron), a multi channelplate LEED (Omicron), a X-ray source (Specs), a differentially pumped He-discharge UV source (Specs), and a hemispherical electron energy analyzer Phoibos 150 (Specs). For recording UPS spectra, a pass energy of 10 eV was used, for the SECO spectra the sample was biased at -10 V and a pass energy of 2 eV was used. XPS spectra were recorded using either Mg Kα (hν = 1253.6 eV) or Al Kα (hν = 1486.6 eV). For satellites and their intensities relative to the main line see Tab. A.5 in

59

Chapter 4. Materials and experimental setups

the appendix. The analyzed spot size was ≈ 2x2 mm^2, while the irradiated area was larger in both cases (UPS and XPS).

In general, for all UPS and XPS experiments performed in this work, photo-induced degradation was systematically checked and no shifts of the spectral features to higher BE during the measurement (reminiscent of sample charging [81]) were observed. The SECO spectra were always measured at least at two positions on the sample to check for homogeneity.

4.3.3. Reflection absorption infrared spectroscopy experiments at Zernike Institute for Advanced Materials

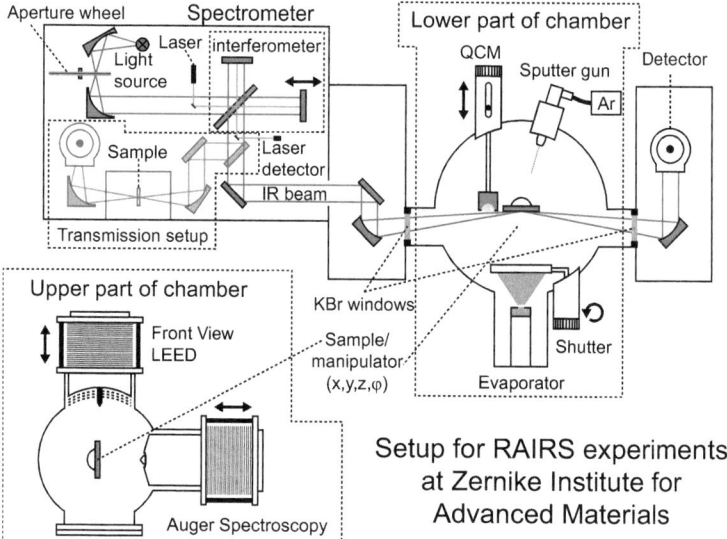

Figure 4.6.: Schematic drawing of the experimental RAIRS setup at the Zernike Insitute for Advanced Materials (top view). The main vacuum chamber consists of an upper and lower part, which can not be separated mechanically. QCM stands for quarz crystal microbalance and LEED for Low energy electron diffraction. The infrared spectrometer can also be used for transmission experiments. The light path for this case is also indicated.

RAIRS experiments presented in this work were performed in the laboratory of Prof. Petra Rudolf at the Zernike Institute for Advanced Materials of the University of Groningen. The experimental setup consisted of a commercial Bruker IFS 66v/S IR spectrometer, which was connected to an UHV chamber as shown in the schematic picture in Fig 4.6. Mid

4.3. Experimental setups

IR radiation was produced by a SiC glowbar source. Before the IR beam traverses the interferometer (containing a beamsplitter made of KBr and a thin metal layer), the spread and shape of the beam were set by a variable aperture. For RAIRS experiments the smallest slit-shaped opening was used with the dimensions 7x1 mm². This produced an irradiated area of approx. 7x8 mm² (irradiated area $\propto$ sec γ·width of the beam) on the sample. Behind the interferometer, the beam was rejoined and passed out of the spectrometer into the UHV chamber through a differentially pumped KBr window. The last mirror focused the beam onto the sample, where it was reflected in specular geometry. The reflected beam is then directed out of the UHV chamber into an external detector housing though another KBr window on the opposite side. Here, the intensity of the beam was detected by a liquid N_2 cooled HgCdTe detector (MCT). To calibrate the interferometer a HeNe laser was coupled into the beam path prior to the interferometer. Behind the interferometer and before the IR beam hits the next mirror the laser beam was outcoupled to the laser detector. To obtain spectra free of atmospheric contaminations, the spectrometer optics and the external detector housing were evacuated (p < 22 mbar).

The UHV chamber (base pressure < 1x10⁻⁹ mbar) had an upper and lower part, which are always connected (Fig. 4.6). The RAIRS experiments were conducted in the lower part, which also houses the organic evaporation sources, a shutter to control the molecular flux onto the sample, a QCM placed directly next to the sample, and an Ar-ion sputter facility to clean the sample. The upper part of the chamber provided a front-view LEED and a combined double pass cylindrical mirror electron energy analyzer and an electron source for Auger electron spectroscopy. The manipulator was coolable to -155 °C and heatable up to 600 °C. The sample temperature could be precisely monitored using a type K thermocouple mounted through a hole in the center of the sample as shown in Fig. 4.7.

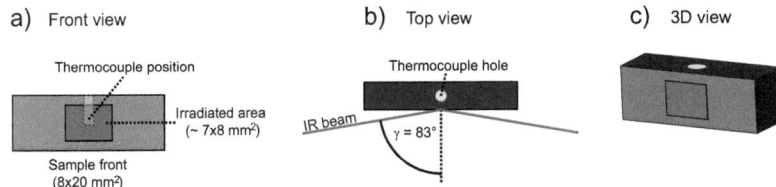

Figure 4.7.: a) front, b) top, and c) 3D view of the sample indicating the irradiated area and the thermocouple position.

The infrared spectrometer could also be used for transmission experiments, for which the beam path is shown in Fig. 4.6, too. The transmission setup was used in this work to obtain reference spectra for the organic materials used in RAIRS experiments.

Chapter 4. Materials and experimental setups

4.4. Experimental details

In this section, the details of the photoelectron and infrared spectroscopy experiments will be illustrated. Furthermore, details of the data evaluation and processing will be shown and estimates for possible errors will be given.

4.4.1. Sample preparation

In general, all experiments start with the cleaning of the specific substrate (as described in Sec. 4.2) on which the measurement is to be performed. The cleaning procedure finishes with a detailed analysis (depending on the techniques available in the respective experimental setup) to check for remaining impurities. In the case of solvent- and ozone-cleaned ITO substrates this step is skipped. Afterwards spectra of the clean substrates are taken, which serve as a reference for subsequently deposited organic materials. For non-metallic ITO substrates a clean polycrystalline gold reference was measured prior to the first measurement to obtain the position of the Fermi level. Usually the evaporation of organic material was done stepwise with the mass thickness θ monitored by the QCM. Consequently all values for coverages correspond to nominal film thicknesses. No correction was made for possible differences in the sticking coefficient between the different substrates and the QCM. Close to an interface (metal/organic or organic/organic) a smaller deposition step size compared to the bulk was used. After each step UPS and/or XPS spectra were recorded and compared to the previous ones in order to monitor the evolution of features with increasing organic film thickness. In the RAIRS experiments spectra were recorded during the deposition with a constant evaporation rate of about 0.3 Å/spectrum. All experiments were performed at room temperature.

4.4.2. UPS and XPS experiments

Most of the UPS and XPS experiments were performed at the SurICat end-station at BESSY II. Additional UPS spectra were recorded at the FLIPPER II end-station at HASYLAB and at the Humboldt-University. After recording, the spectra were divided by the number of sweeps and the effective ring current to normalize them. In the case of the experiments at the Humboldt-University, the spectra were just divided by the number of sweeps. Peak fitting is done by an automated computer routine WINSPEC (University of Namur) using mixed Gaussian and Lorentzian peak shapes and Shirley-type backgrounds. The Fermi level in the valence band (VB) spectra of metallic substrates was obtained by finding the midpoint between the two intersections of the extrapolated base line, the intensity rise at the Fermi level, and the plateau of the s-band states as shown in Fig. 4.4. To convert the spectra from kinetic energy to binding energy (BE), the Au4f 7/2 peak was used as energy reference for

84 eV BE (XPS spectra) or the Fermi level was used as reference for 0 eV BE (UPS spectra). The error of all given values of binding energies and SECO positions is estimated to ±0.05 eV.

4.4.3. RAIRS experiments

Prior to each experiment the position of the sample was calibrated in order to reach the highest intensity at the detector. All scans were taken from 500 cm^{-1} to 3000 cm^{-1} with a resolution of 4 cm^{-1}. For each spectrum 500 scans were co-added to yield a signal-to-noise ratio of better than 10^{-5}. In general both, forward and backward, scans of the interferometer were averaged for one scan. The internal data processing was done using a Blackman-Harris 3-term apodization function, a phase resolution of 128 and a Power-spectrum phase correction. All wavenumber values can be regarded as error free, because they are internally calibrated by the interferogram of the HeNe laser[1]. No baseline correction was applied.

In order to obtain a reference spectrum of the organic material, where all IR active vibrations are allowed and thus observable in the spectrum, samples with the molecular material were analyzed in transmission. The samples were produced by grinding a certain amount of KBr powder with a small amount of the molecular material. A grinding time of 10 min was used, which had proven to be long enough to overcome the Christiansen effect that results in an asymmetric peak shape due to the optically inhomogeneity of the material [129, 130]. Then the grinded powder was filled into a die and pressed into a thin disk of 1 cm diameter by applying the pressure of 12 tons for 2 min. Directly afterwards the disc was put into the spectrometer and a spectrum was taken using the same parameters as described above. The reference for the spectra of the discs was always the blank sample holder. If the adsorption bands were too weak (strong), the amount of molecular material was increased (decreased by the addition of more KBr powder) and pressing a new disc.

4.5. Theoretical calculations

All theoretical calculations of molecules on surfaces presented in this work were performed by O.T. Hofmann and G. Rangger (Technical University of Graz) except when stated otherwise. Here, only a brief description of the concept of density functional theory (DFT) and the further methodology used in the calculations shall be given.

DFT is one of the most widely used computational tools for the prediction of properties of isolated molecules, bulk materials and material interfaces, such as molecules on surfaces. The approach is based on the observation by Hohenberg and Kohn that the ground state

---

[1]The laser has a wavenumber of 15798 cm^{-1}. The wavenumber stability of the unstabilized laser used in the Bruker IFS 66v/S is $\approx \lambda/\Delta\lambda = 10^{-6}$ [128], leading to an error of 0.003 cm^{-1} at a wavenumber of 3000 cm^{-1}.

Chapter 4. Materials and experimental setups

electron density of a system contains in principle all information of the many body-electron wave function [131]. They further showed that the total energy of the system is related to the electron density by a certain functional. If one would know the exact functional the total energy would be minimized, however the exact functional is unknown. Therefore, DFT delicately relies on the choice of functional, which will always be an approximation to the exact one. From these principles Kohn and Sham derived the Kohn and Sham equations [132], which are similar to the Hartree equations (differing only by a exchange correlation potential [133]), but easier to solve reducing the computational effort. They also need to be solved in a self consistently way and at the end one can arrive at the eigenvalues and eigenstates of the system from which all other parameters can be derived. The computational effort roughly increases with the number of atoms of the system N squared [133].

Most of the DFT calculations of the present work were performed using the 3D periodic VASP (Vienna Ab-initio Simulation Package) code [134, 135, 136, 137, 138]. The systems were modeled by the repeated-slab approach, where five layers of metal atoms are used to represent the (111) surface of the metals[2]. The periodicity in z-direction was broken by inserting a sufficiently large amount of vacuum between the slabs (> 20 Å). In the vacuum region, a dipole layer was inserted to prevent artificial polarization of the periodic replicas. For calculations on isolated molecules, a 3D periodically repeated box with the dimensions $20 \times 20 \times 30$ Å^3 was used. In all calculations, the PW91 exchange correlation functional [140] was employed along with the cutoff for the plane wave basis set of 20 Ryd, to expand the Kohn-Sham orbitals of the valence electrons. For the valence-core electron interaction the projector augmented-wave method (PAW) [138, 141] was applied, which allowed for the low kinetic energy cutoff for the plane-wave basis set. The band structure was sampled on a $3 \times 3 \times 1$ Monkhorst-Pack [142] grid of k-points and occupied following a Methfessel-Paxton [143] scheme (broadening: 0.2 eV). All calculations were done in a non spin-polarized manner. The geometry of the individual molecules was obtained by a geometry optimization of a single molecule in a certain unit cell (will be given in the respective results section). Relaxation was performed on a three layer metal slab, where all atoms of the molecule as well as of the uppermost metal layer were fully relaxed using a damped molecular dynamics scheme until the remaining forces were smaller than 0.02 eV/Å. The relaxed geometry was then extended to a five layer metal slab in order to obtain a more reliable description of the electronic structure. Further details are given in Ref. [144].

In addition, gas phase molecular properties and structures were calculated using the program package Gaussian (Gaussian Inc., version 03 [145]). Some of these calculations were done by Georg Heimel (Humboldt University zu Berlin) and the author of the present work. For these calculations, the DFT approach was chosen as well as using the hybrid functional

---
[2]large scale reconstructions, such as the herringbone reconstruction of the clean Au(111) surface[139] were not taken into account.

4.5. Theoretical calculations

B3LYP [146, 147] in conjunction with a 6-31+G* basis set [148]. The neutral molecular systems were allowed to fully relax in their structure until the convergence criteria were met. Frequency calculations were done, to ensure that the found total energy is a real minimum in the potential landscape. The vertical ionization energy (IE) and electron affinity (EA) were obtained by performing single point calculations of the charged radical cation / anion using the fully optimized geometry of the neutral molecular system and taking the difference of their energies according to the following equations:

$$IE_{vert} = E(cation) - E(neutral) \tag{4.1}$$

$$EA_{vert} = E(neutral) - E(anion) \tag{4.2}$$

Furthermore, molecular orbital symmetries and symmetries of vibrational modes for certain molecular point groups were calculated using Gaussian.

Chapter 5.
Results and Discussion

As outlined in chapter 1 the main goal of this work was to characterize the interfaces between conductive electrode surfaces and new electron donor and acceptor materials. As model electrode surfaces, the (111) faces of silver, copper, and gold single crystals were used, since they provide different reactivities and work functions. As application relevant substrate, ITO was used. The initial screening of molecules was done using UPS to obtain the maximum work function modification/change induced by a monolayer of the different molecules. This was done on all three metal single crystals. Using these values as a guide, four molecules were selected for an in-depth characterization using mainly UPS, XPS, and RAIRS. Support for the PES and RAIRS derived conclusions was obtained from DFT modeling and additional experiments, which were performed by collaborators (they will be explicitly indicated, when the results are presented). The structure of the present chapter is as follows:

In the first section (Sec. 5.1), the results of the initial screening process will be presented and trends linking molecular properties with the work function modification potential will be shown. Afterwards, the four molecules that were selected for a detailed characterization will be indicated. These are the electron donor materials MV0 and NMA, and the electron acceptor materials F4-TCNQ and HATCN.

In Sec. 5.2 the adsorption of MV0 on coinage metal surfaces is investigated using PES and RAIRS. It is found that MV0 acts as a strong electron donor on all three substrates and, as a consequence, the work function of the pristine substrates is strongly reduced upon adsorption. The low work function substrates could then be further used to reduce Δ_e into subsequently deposited electron transport materials, such as Alq$_3$ and C$_{60}$. However, MV0 is a rather small molecule (molecular weight: 186 g/mol), exhibiting an evaporation temperature close to room temperature making it unsuitable for idustrial production processes. Consequently, a larger electron donor, namely NMA, is presented in Sec. 5.3. NMA also considerably reduces the work functions of silver, copper, and gold single crystals, although not as strong as MV0. Its structural and electronic properties on the metal substrates are thoroughly characterized by UPS and XPS in the remainder of the section. Furthermore, its properties to lower electron injection barriers into Alq$_3$ will be discussed based on valence band data obtained by UPS.

Chapter 5. Results and Discussion

UPS and RAIRS results for the model acceptor F4-TCNQ on Ag(111) will be shown in Sec. 5.4. They will be discussed with regard to literature results of the adsorption of F4-TCNQ on gold and copper single crystals. Similarly to MV0, F4-TCNQ presents a rather small molecular weight molecule and its tendency to diffuse through subsequently deposited organic layers has already been shown [14, 15, 149]. As a consequence, the larger electron acceptor HATCN is investigated in Sec. 5.5 on Ag(111) in Sec. 5.5 using UPS, RAIRS, scanning tunneling microscopy (STM), LEED, and thermal desorption spectroscopy (TDS). Here, a re-orientation of initially face-on molecules to an edge-on conformation is observed, which is accompanied by a strong work function increase. Finally the high work function template of HATCN on Ag(111) is used to reduce Δ_h into α-NPD. In the next section (Sec. 5.6) the adsorption of HATCN is investigated on Cu(111) and Au(111). In both cases a re-orientation of the initially face-on molecules is found similar to the case on Ag(111). On copper this goes along with a work function increase, while the work function is decreased on gold.

In the last section of this chapter (Sec. 5.7), the adsorption of the strong electron acceptors F4-TCNQ and HATCN on ITO is investigated. A strong increase in work function is found in both cases. The reduction of Δ_h into subsequently deposited α-NPD will be shown and comparisons to the case of F4-TCNQ on ITO will be made. Thereby it is found, that different pinning regimes occur for the same hole transport material (α-NPD). This is reasoned by a different hybridization and the occurrence of a charge transfer type reaction between F4-TCNQ and α-NPD.

5.1. Initial screening of donor & acceptor molecules

In Sec 2.1.4 several properties of metal/organic interfaces have been discussed which collectively determine the energy level alignment. Therefore, it is not possible to judge a-priori if a certain molecular structure will behave as a strong electron donor or acceptor on electrode surfaces. As discussed in Sec. 4.1.1, the Hammet parameter σ of functional groups attached to the molecule has proven to be a first indicator for the electron donating/accepting properties of the molecules. Additional indicators are the DFT calculated gas phase IE and EA values of the molecules, which also serve as a first benchmark before actual synthesis. Nevertheless, it is still necessary to screen the synthesized molecules experimentally, to validate and quantify the work function modification. In this work, the molecules have been initially screened by UPS on the (111) faces of gold, copper, and silver metal single crystals. The obtained work function modifications of saturated molecular layers ($\Delta\Phi$) are plotted in Fig. 5.1 versus the work function of the clean substrates (Φ_S). One trend can clearly be observed: The maximum positive work function modification increases while the negative one decreases with decreasing

5.1. Initial screening of donor & acceptor molecules

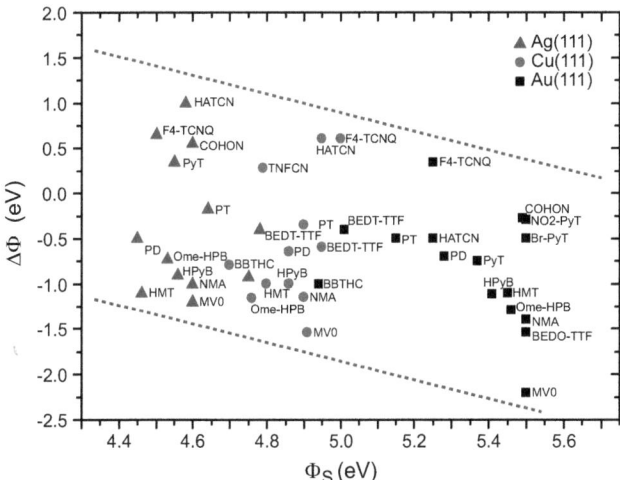

Figure 5.1.: Saturated work function changes $\Delta\Phi$ as a function of initial substrate work function Φ_S for the molecules screened in this work adsorbed on Au(111), Cu(111), and Ag(111). The values for F4-TCNQ on gold and copper have been taken from Refs. [9, 16].

substrate work function Φ_S. The spread in work function modification $\Delta\Phi$ is ca. 2.5 eV, or in other words: The work function of the substrate metals can be tuned within a range of ca. 2.5 eV by the molecules screened in this work. Note however, that the range is shifted depending on the substrate's work function Φ_S. A simple explanation for the dependence of the maximum positive (negative) work function modification on the substrate's work function is the energetic difference between the substrate's Fermi level and the LUMO (HOMO) of the molecule. In the hypothetic case of vacuum level alignment, the LUMO (HOMO) would be located closer to the Fermi level for low (high) work function materials, which would make a larger charge transfer and thus a higher work function modification possible. The same would be expected if the work function is kept constant and the molecular levels (i.e. the HOMO and LUMO) are varied. That this trend is indeed observed is shown in Fig. 5.2 and 5.3, where the work function modification $\Delta\Phi$ is plotted versus the DFT calculated vertical IE_{theo} and EA_{theo}. A roughly linear relationship is obtained in both cases, indicating that the larger the IE_{theo} and EA_{theo} of the molecules are the more the properties of the molecules change from electron donor to acceptor. This also shows that these values can serve as an indicator for the molecular properties. In all cases, where the ionization energy was experimentally accessible, a linear dependency between IE_{exp} and IE_{theo} was found, which deviated only slightly from

Chapter 5. Results and Discussion

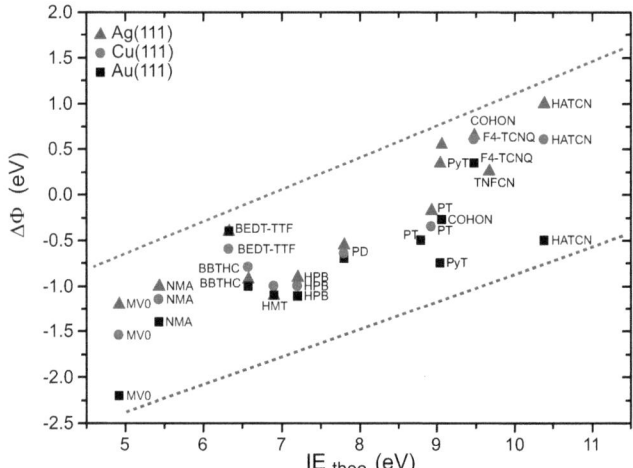

Figure 5.2.: Saturated work function changes $\Delta\Phi$ as a function of DFT calculated (vertical) ionization energy IE_{theo} for the molecules screened in this work adsorbed on Au(111), Cu(111), and Ag(111). The red dashed lines are intended as a guide to the eye. The values for F4-TCNQ on gold and copper have been taken from Refs. [9, 16]. The IE_{theo} values were calculated by Oliver T. Hofmann (TU Graz, Austria).

a linear with slope 1. This underlines the qualitative accuracy of DFT calculated IE_{theo} values. The highest negative work function modifications were achieved for MV0 on all three substrates. Consequently, this molecule was chosen for an in-depth study. In addition, NMA was selected, because it exhibits strong similarities in its chemical structure compared to MV0 and has a much higher molecular weight. Its performance to lower the work function is also good on all three metals. The highest positive work function modifications were achieved with F4-TCNQ and HATCN on low work function metals. They are also similar in molecular structure and thus make a comparative study possible.

5.2. Work function reduction of up to 2.2 eV with an air-stable molecular donor layer of MV0

5.2.1. Introduction

In this section, MV0 is analyzed with regard to its electron donation properties. From the literature it is known, that MV0 acts as a strong electron donor in charge transfer compounds

5.2. MV0 on coinage metals

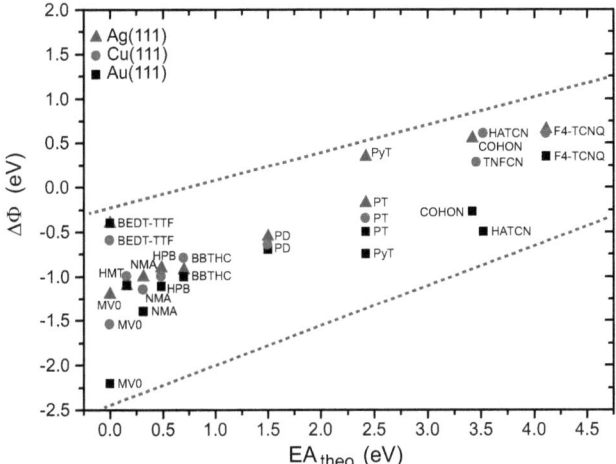

Figure 5.3.: Saturated work function changes $\Delta\Phi$ as a function of DFT calculated (vertical) electron affinity EA_{theo} for the molecules screened in this work adsorbed on Au(111), Cu(111), and Ag(111). The red dashed lines are intended as a guide to the eye. The values for F4-TCNQ on gold and copper have been taken from Refs. [9, 16]. The EA_{theo} values were calculated by Oliver T. Hofmann (TU Graz, Austria).

[150]. Here, UPS and XPS are used to investigate the adsorption of MV0 on the (111) faces of silver, copper, and gold single crystals. As a result of molecule-to-metal electron transfer, the work function of all three substrates is strongly reduced to $\Phi_{mod} = 3.35$ eV. This is reminiscent of Fermi level pinning, which is further supported by the position of the valence molecular orbitals. RAIRS is used to analyze the adsorption geometry and ionicity of MV0 adsorbed on the Ag(111) surface revealing that the adsorbed species has adsorption bands most similar to MV+1, thus corroborating the electron donation found by UPS. In addition, the energy levels of the ETMs (Alq$_3$ and C$_{60}$) deposited on top of MV0 modified Ag and Au surfaces are investigated in order to test if MV0 interlayers can be used to reduce the electron injection barriers. It is found that the molecular energy levels are shifted to higher BE compared to layers on pristine Ag and Au. Assuming a constant transport gap, these results show that Δ_e is largely reduced. The effect is strongest for Alq$_3$ deposited on MV0 pre-covered Au(111) surfaces, where Δ_e is reduced by 0.8 eV. Parts of this section have been published in Ref. [151].

Chapter 5. Results and Discussion

5.2.2. Valence electronic structure

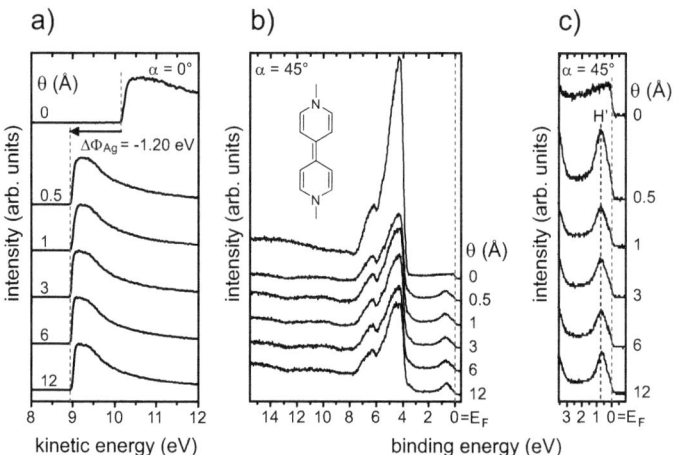

Figure 5.4.: The valence electronic structure of the sequential deposition of MV0 on Ag(111) as obtained by UPS is shown. a) SECO spectra, b) full valence band spectra, and c) zoom into the near Fermi level region. H' denotes the metal-molecule hybrid state derived from the HOMO of neutral MV0 and E_F the Fermi level. The inset in b) shows the molecular structure of MV0.

MV0 was deposited sequentially on coinage metal single crystals. After each deposition step, valence band and SECO spectra have been taken, which are shown in Figs. 5.4 (silver), 5.5 (copper), and 5.6 (gold). The saturation coverage is reached on all three substrates already with the very first deposition of 0.5 Å. Two possibilities might explain this behavior: (i) the sticking coefficient of MV0 on the QCM is much lower than on the clean surfaces or (ii) the molecules build a saturated vapor with a certain partial pressure (during the experiments shown in Figs. 5.4, 5.5, and 5.6, the pressure in the chamber was $p = 1 \cdot 10^{-7}$ mbar) in the chamber and thus adsorption from the vapor phase might result. However, (i) can be clearly ruled out, because deposition was also possible with the sample not directed towards the molecular source. (ii) can be substantiated as follows: If the evaporation of MV0 was done very quickly with a lower pressure ($p = 5 \cdot 10^{-8}$ mbar)[1], the evolution of the adsorption on Au(111) could be followed as shown in Fig. 5.7b for the SECO spectra. Noteworthy, a longer time t was needed to reach the full saturated coverage at this pressure. At the

---

[1] The base pressure in the chamber was the same ($p = 1 \cdot 10^{-8}$ mbar), however, the current used to heat the MV0 evaporator was reduced compared to the previous case.

5.2. MV0 on coinage metals

saturation coverage, only a very thin film (which will be referred to as "saturated monolayer coverage" in the course of this work) was adsorbed on all three substrates, which is evident from the weak attenuation of the Fermi-edge in Figs. 5.4c, 5.5c, and 5.6c. Consequently, it must be concluded that multilayers are thermodynamically unstable and desorb in a time range equivalent to the sample transfer time (transfer from the evaporation chamber into the measuring position $\approx$ 5 min). In addition the peak areas of the core levels remain constant for depositions beyond the saturated monolayer. These results are in good agreement with the adsorption of 4,4'-bipyridine on TiO_2 from the vapor phase, which also saturates after completion of *one* monolayer [152].

The SECO spectra in Fig. 5.4a evidence that upon MV0 deposition on Ag(111), Φ_{Ag} decreases from 4.60 eV (pristine Ag(111)) to 3.40 eV at the saturated monolayer coverage of MV0 on Ag. This change of $\Delta\Phi = -1.20$ eV was already reached with the first deposition as shown in Fig. 5.4a and remained constant for higher nominal coverages (up to 12 Å). Valence band (VB) spectra for the MV0 coverages corresponding to the ones shown in the SECO spectra, are plotted for a take-off angle $\alpha = 45°$ in Figs. 5.4b and c. The overall attenuation of Ag(111) photoemission features (especially the Ag 4d level) is accompanied by the emission of molecular derived features. These are rather broad and low in intensity except for the HOMO at 0.75 eV BE (denoted H' in Fig. 5.4c), which is well resolved. The HOMO is located right at the Fermi-edge, however, it is very symmetric and only slightly (if at all) cut by Fermi level. Upon subsequent deposition, the HOMO intensity is minimally decreased, however its position remains constant. Higher BE molecular features appear as shoulders left and right of the Ag 4d level and between 9 and 12 eV BE and do not shift in position either. Spectra taken at $\alpha = 0°$ (normal emission) indicate that the intensity of the broad feature between 9 and 12 eV BE (denoted A in Fig. 5.7a, with a peak maximum at 10.15 eV BE) and of the HOMO are dependent on α; the former increases in intensity whereas the latter decreases. The peak positions, however, remain constant for different take-off angles.

The SECO and VB spectra of the sequential deposition of MV0 on Cu(111) show a qualitatively similar picture compared to Ag(111) as plotted in Fig. 5.5. The work function of the pristine substrate ($\Phi_{Cu} = 4.90$ eV) decreases by 1.55 eV for the saturated monolayer of MV0 ($\Phi_{Cu,MV0} = 3.35$ eV) as shown in Fig. 5.5a. The VB spectra in Fig. 5.5b and c taken at $\alpha = 45°$ show the attenuation of Cu photoemission features (especially the Cu 3d levels), while MV0 derived features rise in intensity. Among these, the HOMO peak maximum (denoted H' in Fig. 5.5c) is located at 0.75 eV BE for all nominal coverages (up to 24 Å). In full analogy to the results on Ag(111), it does not seem to be cut by the Fermi level. Its intensity remains constant for all spectra. Furthermore, between 6 and 12 eV BE *two* broad molecular features are observed. Their appearance is confirmed in the spectra taken at $\alpha = 0°$ (Fig. 5.7a), where their intensity is increased and clearly two separate peaks can

Chapter 5. Results and Discussion

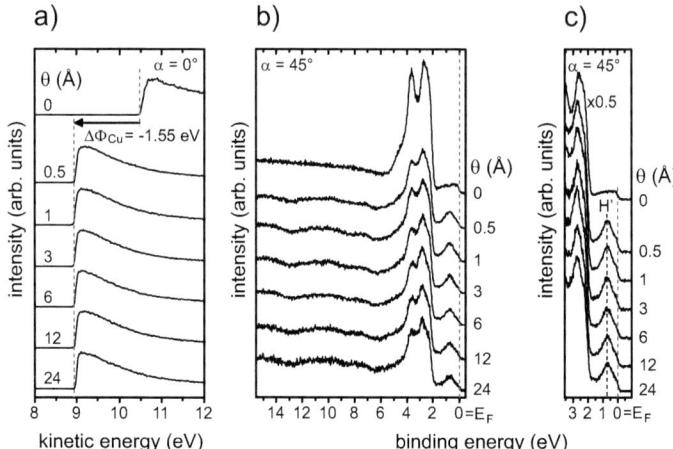

Figure 5.5.: The valence electronic structure of the sequential deposition of MV0 on Cu(111) as obtained by UPS is shown. a) SECO spectra, b) full valence band spectra, and c) zoom into the near Fermi level region. H' denotes the metal-molecule hybrid state derived from the HOMO of neutral MV0 and E_F the Fermi level.

be discerned (denoted A and B in Fig. 5.7. Their peak maxima are located at 10.15 eV BE (A) and 7.70 eV BE (B)).

The saturated monolayer of MV0 on Au(111) was reached already after the deposition of nominally 0.5 Å at a pressure of $p = 1 \cdot 10^{-7}$ mbar (see above). The SECO spectra of this experiment are shown in Fig. 5.6a. Here, the work function of pristine Au(111) ($\Phi_{Au} = 5.50$ eV) decreases by 2.20 eV to a value of $\Phi_{Au,MV0} = 3.30$ eV for the saturated monolayer of MV0. When the current used to heat the MV0 source was reduced, which resulted in a lower pressure ($p = 5 \cdot 10^{-8}$ mbar), and the exposure time t was reduced to some seconds, the evolution of the work function could be followed as shown in Fig. 5.7b and c. Note that, the reduction of Φ is not linear with increased exposure time, which points to either a nonlinear growth process (as observed for alkanethiols adsorbed from the vapor phase [153]) or strong depolarization with increasing dipole density [29]. The VB spectra taken at $\alpha = 45°$ are plotted in Fig. 5.6b and c. They show an overall attenuation of gold photoemission features together with the emergence of molecular features. In contrast to VB spectra of MV0 on Ag(111) and Cu(111), no intensity close to the Fermi-edge is observed in the spectra on gold irrespective of the take-off angle α (data not shown). At higher BE, feature A with its peak maximum at 10.15 eV BE is observed, which shows the established angular intensity dependence (compare

5.2. MV0 on coinage metals

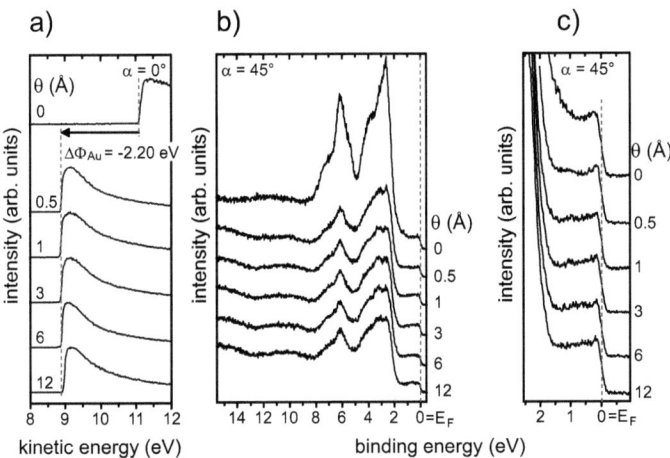

Figure 5.6.: The valence electronic structure of the sequential deposition of MV0 on Au(111) as obtained by UPS is shown. a) SECO spectra, b) full valence band spectra, and c) zoom into the near Fermi level region. E_F denotes the Fermi level.

Fig. 5.6b ($\alpha = 45°$) and Fig. 5.7a ($\alpha = 0°$)). Its peak position and intensity remains constant with further deposition, similar to the results on silver and copper.

The angular dependence of the HOMO feature in the VB is shown in Fig. 5.7 and in more detail in Fig. 5.8a. In normal emission ($\alpha = 0°$) the intensity of the HOMO feature of the saturated monolayer of MV0 on Ag(111) (Fig. 5.8a) is much lower compared to the spectra taken at off-normal emission ($\alpha = 45°$). This can be explained in first approximation by a different number of allowed transitions from an initial to a final state in the PES experiment. The initial state is the HOMO which has been computed for MV0 and MV+1 using the Gaussian software and DFT methodology (done by Oliver T. Hofmann, TU Graz, Austria; details see Sec. 4.5). The orbital population is shown in Fig. 5.8c and d and it is exclusively located on p_z orbitals for both neutral and ionic states. MV0 and MV+1 belong to the symmetry point group C_{2v}, which still holds for the adsorbed species as the mirror plane parallel to the plane of the aromatic rings is already lost in the gas phase molecule. This is in agreement with theoretical results of the parent molecule viologen[2] adsorbed on Ag(111) [154]. The HOMO has the irreducible representation A_1 and is thus totally symmetric. As outlined in Sec. 3.1.3, the final state either belongs to the A_1 (normal emission) or to a representation, which is even for the mirror plane spanned by the incident light and the

---

[2]The methyl groups are substituted by hydrogen atoms in viologen compared to methyl viologen.

Chapter 5. Results and Discussion

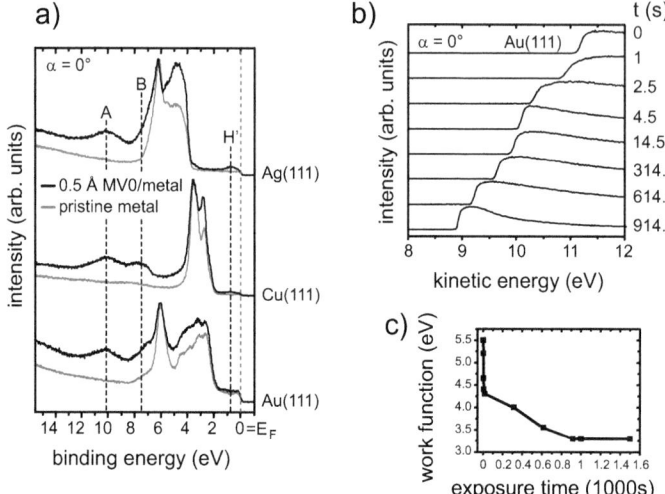

Figure 5.7.: a) Valence band spectra of pristine (blue) and MV0 (saturated monolayer, black) covered Ag(111), Cu(111), and Au(111) substrates obtained at normal emission. H' denotes a metal-molecule hybrid state derived from the HOMO, A and B deeper lying molecular orbitals, and E_F the Fermi level. b) SECO spectra of the adsorption of MV0 on Au(111) done at a pressure of $p = 5 \cdot 10^{-8}$ mbar with increasing exposure time (t). c) Evolution of the work function as obtained from the spectra in b) plotted versus the t.

detector (off-normal detection). These are either B_1 or B_2 depending on the orientation of the molecular axis with respect to the mirror plane. As shown in Sec. 3.1.3 the product of $\vec{A}_1$ and $\underline{P}$ yields only y and z components, which holds for the present experiments, since polarized synchrotron radiation was used. Thus three direct products need to be evaluated for normal and off-normal emission:

normal emission:

$$final\,state \otimes momentum\,operator \otimes initial\,state = result$$

$$A_1 \otimes A_1 \otimes A_1 = A_1 \quad allowed$$

5.2. MV0 on coinage metals

$$A_1 \otimes B_1 \otimes A_1 = B_1 \quad \text{not allowed}$$

$$A_1 \otimes B_2 \otimes A_1 = B_2 \quad \text{not allowed}$$

off-normal emission:

$$A_1 \otimes A_1 \otimes A_1 = A_1 \quad \text{allowed}$$

$$B_1 \otimes B_1 \otimes A_1 = A_1 \quad \text{allowed}$$

$$B_2 \otimes B_2 \otimes A_1 = A_1 \quad \text{allowed}$$

In both cases the first transition containing the totally symmetric representation A_1 is allowed. However, the other two transitions are forbidden for normal, but allowed for off-normal detection. In a first order approximation, this should result in a higher intensity in the off-normal case, since there more transitions are allowed. Since both measurements were carried out on the same sample, the rotation of the molecules on the surface and thus the contribution from light with symmetry A_1, B_1 and B_2 should be the same. From this result it can be deduced that the molecules are oriented face-on with respect to the substrate surface, because edge-on orientation as shown in Fig. 5.11 would result in C_S surface symmetry. This would lead to the same number of allowed transitions for normal and off-normal detection (In C_S symmetry only one mirror plane exists and any operator being even with respect to this mirror plane belongs to the totally symmetric representation. Consequently, in both cases two transitions are allowed). Since the same angular dependence of the HOMO is found on copper, as shown in Fig. 5.8b, the same arguments hold here. These results give a first indication that the molecules are oriented face-on with respect to the substrate surface on Ag(111) and Cu(111) and present the motivation for the RAIRS experiments shown in Sec. 5.2.4.

The energy level alignment of MV0 on the three coinage metals is summarized in the schematic energy diagram in Fig. 5.9. On Ag(111) the work function reduction of $\Delta\Phi_{Ag,MV0} = -1.20$ eV is smallest compared to the two other metals, where the work function is decreased by $\Delta\Phi_{Cu,MV0} = -1.55$ eV (Cu(111)), and $\Delta\Phi_{Au,MV0} = -2.20$ eV (Au(111)). The modified work function of the metals due to the saturated monolayer of MV0 is, however,

Chapter 5. Results and Discussion

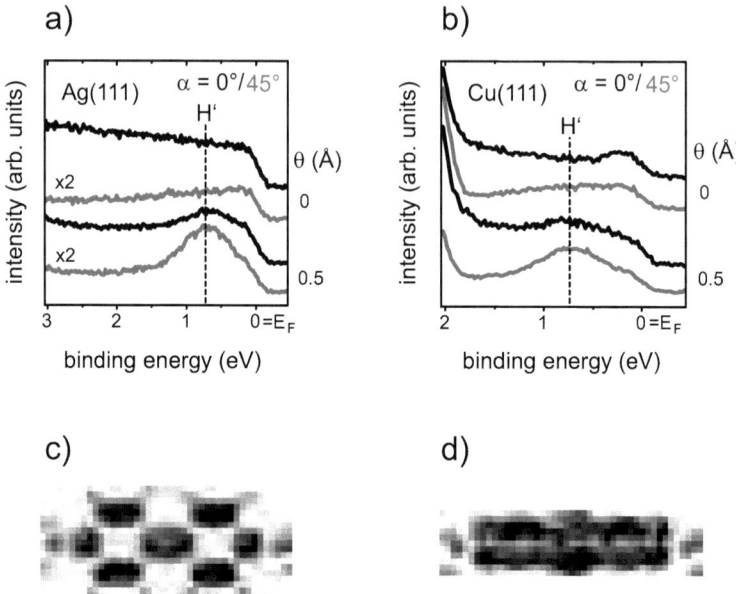

Figure 5.8.: *Upper part:* Zoom into the valence band region close to the Fermi-edge for pristine and MV0 covered Ag(111) (a) and Cu(111) (b) for normal (black curves) and off-normal ($\alpha = 45°$, red curves) electron detection. H' denotes the metal-molecule hybrid state derived from the HOMO of neutral MV0 and E_F the Fermi-edge. *Lower part:* Isodensity representation of the HOMO of MV0/MV+1. c) side and d) top view. The isodensity values were set to 0.004 e·Å^{-3}.

the same in all three cases with a value of $\Phi_{mod} = 3.35$ eV. This is reminiscent of Fermi level pinning [49, 155], which is further evidenced by the close proximity of the HOMO to the Fermi-edge on Ag(111) and Cu(111). This must also apply for MV0 on Au(111), because peak A is found at the same position with respect to the Fermi level on all three substrates, even though the HOMO has not been resolved spectroscopically here. Fermi level pinning occurs, when the HOMO position of the isolated molecule would be situated above the Fermi level of the metal (due to its high work function compared to the molecular IE) in the combined system (molecule on metal). Since this is thermodynamically unfavorable, charges will flow *from the molecule to the metal* to bring the HOMO below the Fermi level and the whole system in thermodynamic equilibrium. This is the same for all three substrates, however, the details of the interaction between MV0 and the substrate might still be different for the three metals.

5.2. MV0 on coinage metals

Thus, the hybridization between HOMO and substrate levels can be different on Au(111) compared to Ag(111) and Cu(111) resulting in a much smaller photoemission intensity at the tested energies and angles. Charge transfer is also supported by the reduction of the work function (far) below the values reported for purely physisorbed systems such as Xenon or TTC on Ag(111) (-0.5 eV), Cu(111) (-0.5 eV), and Au(111) (-0.7 eV).

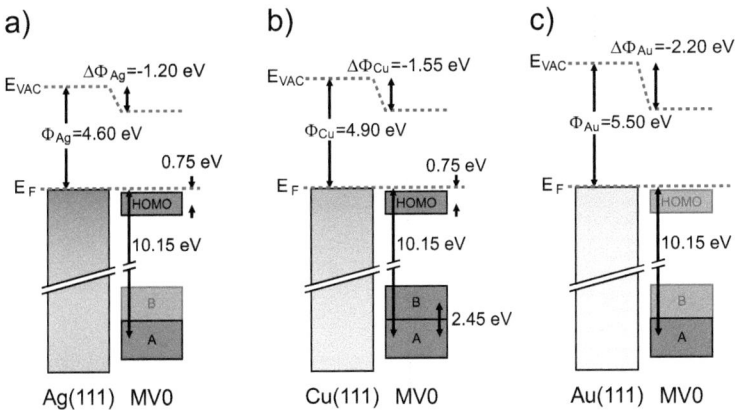

Figure 5.9.: Schematic energy level diagrams for the saturared monolayer of MV0 adsorbed on Ag(111), Cu(111), and Au(111). E_F denotes the Fermi level, E_{VAC} the vacuum level, and A and B the deeper lying molecular orbitals. Orbitals colored in light green are not observed in the photoemission spectra.

5.2.3. Core level analysis

The N1s core level spectra of a saturated monolayer of MV0 on Ag(111), Cu(111), and Au(111) are shown in Fig. 5.10. The spectrum of the adsorption of MV0 on gold shows two peaks. They are located at 400.1 eV (N1) and 401.2 eV BE (N2) and their intensity ratio is approximately 1:1. The FWHM of both peaks is 0.9 eV. In contrast, only one single peak is found in the case of MV0 on Ag(111) and Cu(111), which is located at the position of the N1 peak. The FWHM of both peaks is 0.75 eV and thus even smaller compared to the N1s peaks on gold, precluding the existence of two overlapping peaks on silver and copper. Therefore, all nitrogen atoms of MV0 adsorbed on Ag(111) and Cu(111) must have the same chemical environment. On Au(111) the N2 peak must belong to a nitrogen species that is more positively charged compared to the N1 species, since higher BE means lower kinetic energy of the photoelectron due to the stronger attraction by the positive core (the charge is less effectively screened in this case, see Sec. 3.1.4). Corroborating the findings

Chapter 5. Results and Discussion

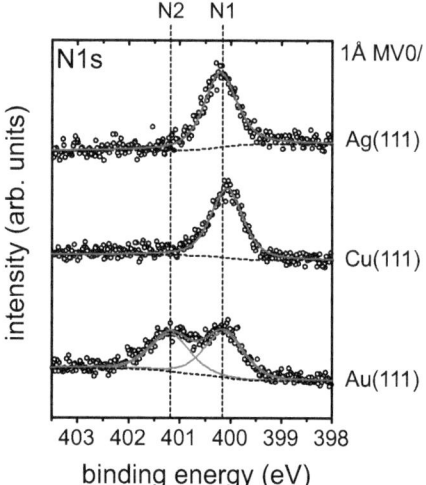

Figure 5.10.: Nitrogen core level spectra of a saturated monolayer of MV0 on Ag(111), Cu(111), and Au(111). The spectral deconvolution has been done using mixed Gaussian and Lorentzian peaks and Shirley backgrounds.

from the N1s core levels, the spectral shape of the C1s core level is similar on copper and silver, but different to the recorded ones on gold (see Fig. A.2 in the appendix). However, the spectral assignment of the different carbon peaks is rather difficult, since the exact details of the interaction and the surface orientation are unknown, which will both affect the spectral shape of the core levels [9, 152]. From the N1s spectra, where only one single symmetric peak with a small FWHM is observed for MV0 adsorbed on Ag(111) and Cu(111), it must be concluded, that the surface adsorption geometry of MV0 is symmetric with respect to the nitrogen atoms. Otherwise one nitrogen spectra should show a (small) chemical shift, resulting in two peaks, a shoulder or at least broadened peaks. Two extreme orientations can be envisioned, in which the long molecular axis is parallel to the surface and consequently the nitrogen atoms are in a symmetric position: face-on and edge-on as shown in Fig. 5.11b. The stoichiometric ratio of carbon and nitrogen in MV0 is 6:1. If the peak areas (metal core level, C1s, N1s) are integrated, corrected with the respective ASF values, and normalized to the total sum as described in Sec. 3.1, a similar amount of about 30-35 % MV0 per surface area is obtained on all three substrates. The ratio between carbon and nitrogen on the surfaces is 5.2:1 (Ag(111)), 5.1:1 (Cu(111)), and 6.4:1 (Au(111)). These values are in good agreement with the stoichiometric ratio and the slight deviation may be due to the use of polarized light and a different beam energy compared to the unpolarized Al Kα radiation used for deriving the ASF values.

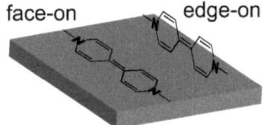

Figure 5.11.: Two possible orientations of MV0 adsorbed on a solid surfaces, where both nitrogen atoms are in an equivalent position, denoted as face-on and edge-on.

5.2.4. RAIRS experiments of MV0 on Ag(111)

To obtain more insights into the surface orientation and ionicity of MV0 adsorbed on Ag(111), the evolution of vibrational modes for the adsorption has been studied *in-situ* using RAIRS. In literature several infrared studies have been carried out for MV+2 (bi-cation of MV0) in KCl solution embedded in an electrochemical cell [156, 157, 158, 159]. With the setups used in these studies, it was possible to obtain the different redox species MV0 (i.e. MV+1, and MV+2) by simply changing the applied potential. However, although in all cases reflection setups were used only Christensen and coworkers [158] where able to distinguish between adsorbed species (the surface selection rule holds) and species formed in solution above the surface (the surface selection rule does *not* hold). In the present work, the experiments were carried out on the Ag(111) surface and thus the surface selection rule holds. The reference spectrum of the clean Ag(111) substrate was taken at a sample temperature of 170 °C, because at this temperature no MV0 adsorption from the vapor phase was observed. When the sample was allowed to cool down towards room temperature, a gradual adsorption of MV0 on Ag(111) was observed as shown in Fig. 5.12. The spectra contain several modes, which are observed to shift during adsorption. For all modes the shift occurs to higher wavenumbers, except for the mode at 742 cm^{-1}, which shifts to *lower* wavenumbers. This mode is assigned to out-of-plane vibrations of the hydrogen atoms bonded to the azine rings (see inset in Fig. 5.12) in accordance to literature [114, 156]. As explained in Sec. 3.2.3 several effects can be responsible for the frequency shifts of vibrational modes of adsorbed species, when their coverage is increased. For several molecules it has been found, that the frequency of out-of-plane vibrations of hydrogen atoms bonded to aromatic or non-aromatic systems decreases with increasing coverage [114, 160, 161]. This was attributed to an increased repulsive interaction [161]. A shift to higher wavenumbers with increasing coverage (as observed for all other modes) can result from dipole-dipole coupling of the adsorbed species (dynamical shift), which becomes stronger, when the molecules are more densely packed [106]. However,

Chapter 5. Results and Discussion

also chemical shifts can occur, if for example the CT between adsorbed species and substrate is increased or decreased as described by Blyholder [117]. Note that it is also possible that both effects occur at the same time, and shifts in opposite directions result, which partially cancel each other. As a result only a small shift may be observed. As MV0 on Ag(111) undergoes a strong CT with the substrate and a strong dipole across the interface is established, it cannot be resolved at present, which contribution (dynamical or static shift) dominates the shifts of the modes above 1000 cm^{-1} (5.12). These modes belong to in-plane stretching and deformation vibrations of the aromatic system, except for the mode at 1416 cm^{-1}, which is attributed to C-H deformation vibrations of the methyl groups [158]. This may also explain why this mode is shifted only by 7 cm^{-1} compared to the shift of about 20 cm^{-1} of the other modes (above 1000 cm^{-1}), since it is not part of the aromatic system and thus not as strongly influenced by the CT as the other vibrations. After the last spectrum, which is denoted "saturated monolayer", no further increase or shift of the vibrational bands was observed, even when the temperature of the molecular source and thus the MV0 vapor pressure in the chamber was increased. The number of observed modes and their positions in the spectrum of the saturated monolayer correspond to the ones observed by Christensen and coworkers [158] for surface adsorbed MV+1 (except for the mode at 742 cm^{-1}, because their setup did simply not allow for the detection of modes lower than 800 cm^{-1}). This is a strong indication that MV0 donates charges to the Ag(111) substrate and becomes oxidized to MV+1. The most intense mode in the spectrum of the saturated monolayer is the out-of-plane vibration of the hydrogen atoms bonded to the azine rings at 742 cm^{-1}. Besides, only a small intensity is observed for the stretching vibration of the hydrogen atoms of the methyl group (not shown). Consequently, it can be concluded that the molecules are oriented with their long and short molecular axis (almost) parallel to the substrate surface (face-on in Fig. 5.11). The observed in-plane stretching vibrations of the aromatic system are IR allowed, because they create a dynamical dipole perpendicular to the surface as explained in Sec. 3.2.

5.2.5. DFT results

The following theoretical results have been obtained by Oliver T. Hofmann (TU Graz, Austria).

Molecular properties such as IE and EA of a single MV0 molecule in the gas phase were obtained using the Gaussian software as described in Sec. 4.5. Both computed vertical values are exceptionally low ($IE_{MV0}^{vert} = 4.92$ eV and $EA_{MV0}^{vert} \approx 0$ eV). The theoretical description of MV0 adsorbed on metal surfaces using the VASP code (as outlined in Sec. 4.5) was done for face-on molecules. Since it was not possible to obtain structural insights of the ordering of MV0 on any substrate, a $5 \times 3\sqrt{3}$ unit cell was used in the calculations. A model containing face-on molecules was also used for Au(111), since the detailed structure could not be obtained

5.2. MV0 on coinage metals

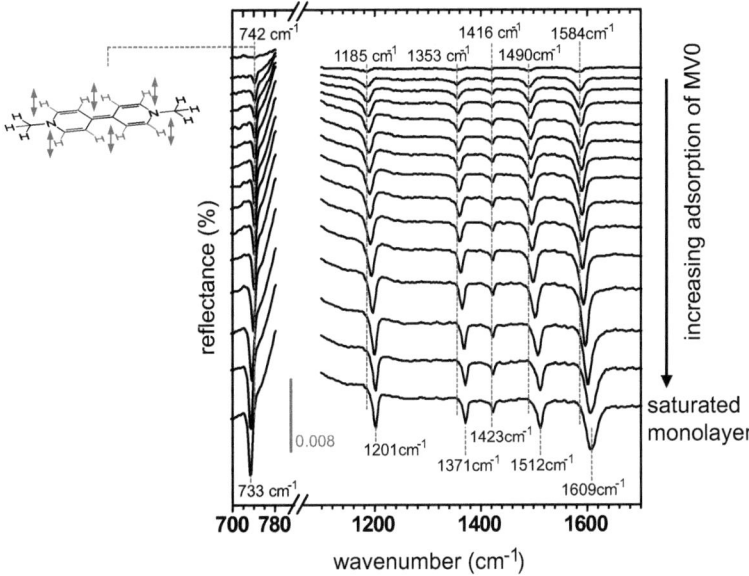

Figure 5.12.: RAIR spectra recorded during the adsorption of MV0 on Ag(111) while the sample was allowed to cool down from 170 °C (where the reference was obtained) to room temperature. For longer exposure time, no change in the spectra corresponding to the saturated monolayer was observed. The inset shows the out-of-plane vibration of the peripheral hydrogen atoms.

experimentally as described before. The density of states for the gas phase molecule is shown in Fig. 5.13.

The modeling enables to rationalize the origin of the Φ lowering by MV0 deposition: For the MV0 covered Au surface, the Fermi level of the system cuts through the HOMO derived peak of the density of states (DOS) as shown in Fig. 5.13, indicative of a significant electron transfer from the molecular HOMO derived band to the Au substrate. From integration over the plane averaged charge rearrangements, a transfer of ≈ 0.8 electrons from the molecule to the metal can be inferred, which results in the formation of an interface dipole. This dipole leads to a calculated reduction of the system's Φ by -1.40 eV. The smaller value compared to the experimental observation is most likely a consequence of the unknown structure of MV0 on the Au(111) surface and the low coverage of the monolayer used in the calculations. For example, if a more tightly packed layer with one MV0 molecule in a $3 \times 3\sqrt{3}$ unit cell is used, then the Φ modification is -1.9 eV. The theoretical results obtained for MV0 on Au(111) are

83

Chapter 5. Results and Discussion

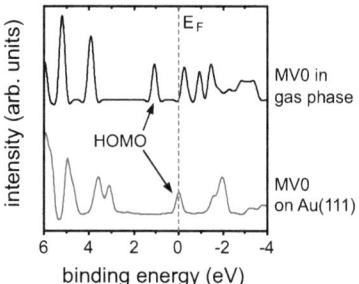

Figure 5.13.: DFT calculated DOS for MV0 in the gas phase and adsorbed on Au(111). E_F denotes the Fermi level.

in line with the results of the parent molecule viologen on the same substrate [154]. Since the methyl groups have only a minor influence on the aromatic system [120], the results found for viologen should (with some care) be applicable to the case of MV0. For viologen it was found that the HOMO of the adsorbed molecule was only occupied by 60% compared to the molecule in the gas phase, as a result of the molecule to metal electron donation. The adsorption geometry was found to be slightly bend for the adsorption on Au(111). Both, the charge transfer and the bend of the molecule, resulted in a change from a quinoidal to benzoidic geometry of MV0. This stabilizes the charge and can have a signifcant influence as driving force for the transfer [162]. Similar results have been found for viologen adsorbed on Ag(111) and Cu(111), however, the HOMO occupation was slightly higher meaning that less charge was transfered. In all three cases the work function of the substrate was significantly reduced: $\Delta\Phi_{Au,viologen} = -1.21$ eV, $\Delta\Phi_{Ag,viologen} = -0.82$ eV, and $\Delta\Phi_{Cu,viologen} = -1.05$ eV, which is in qualitative agreement with experimentally observed values for MV0. Thus, it can be infered that similar arguments as the aromatic stabilization upon charge donation play an important role in the adsorption of MV0 on coinage metal surfaces as well.

5.2.6. Electron injection barrier tuning with MV0 interlayers

To test the effect of a Φ-reducing MV0 interlayer on the charge injection barriers at interfaces between prototypical ETMs and electrode surfaces, Alq$_3$ and C$_{60}$ were deposited on saturated layers of MV0 (Φ_{mod} at a minimum) on Ag(111) and Au(111) substrates. Note that it could be possible that the saturated monolayer of MV0 on the surfaces is not a fully closed layer and thus holes might exist where the ETMs can reach the metal surface directly. The results of the UPS experiments are plotted in Figs. 5.14 - 5.17. They show that the molecular peaks of the ETMs have a similar FWHM for both cases of MV0 modified and clean metal

5.2. MV0 on coinage metals

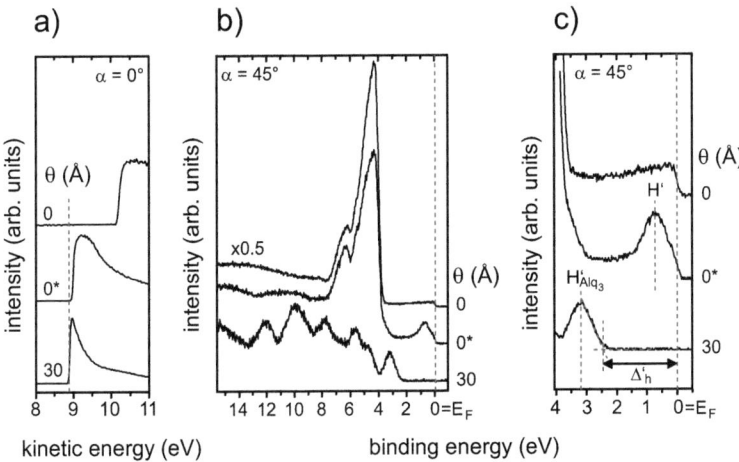

Figure 5.14.: Valence band and SECO spectra for the deposition of Alq$_3$ on MV0 pre-covered Ag(111). The thickness of the Alq$_3$ film was 30 Å. The HOMO position of MV0 is labeled H', that of Alq$_3$ H'_{Alq_3}, its onset by Δ'_h. The MV0 pre-covered spectrum is denoted by "0*" and the Fermi level by E_F.

surface. This indicates that most of the ETM molecules are aligned with the same (modified) work function (as in the other case a significant broadening would result from the different local work functions). In general three to four layers of Alq$_3$ or C$_{60}$ were needed, to obtain the HOMO position of the bulk (i.e. without the influence of the screening induced by the substrate). Fig. 5.14a shows that the work function change induced by the saturated MV0 monolayer ($\Delta\Phi_{Ag,MV0} = -1.2$ eV, spectrum denoted as "0*") is retained upon 30 Å Alq$_3$ deposition (spectrum denoted as "30"). This indicates, that vacuum level alignment occurs between MV0 and Alq$_3$, meaning that no new dipole is created at this interface and the work function of the system remains at $\Phi_{mod} = 3.30$ eV. The molecular levels of Alq$_3$ align rigidly to this new vacuum level as shown in Fig. 5.14b and c. Consequently, the HOMO peak maximum of Alq$_3$ (denoted H'$_{Alq3}$) is located at 3.20 eV BE with its onset at $\Delta'_h = 2.50$ eV BE. In contrast, the deposition of Alq$_3$ with a similar thickness on pristine Ag(111), yields a HOMO onset of 2.40 eV [163]. The vacuum level has been reported to be reduced by 1.10 eV relative to pristine Ag(111). Thus the insertion of the MV0 interlayer is responsible for lowering the vacuum level beyond the value of Alq$_3$ on pristine Ag(111) by 0.1 eV. Consequently, the HOMO onset is also shifted by this amount to higher BE. By subtracting the HOMO onset from the transport gap of Alq$_3$ $E_t = 4.6 \pm 0.4$ eV [24] the

Chapter 5. Results and Discussion

electron injection barrier can be estimated. This yields $\Delta_e = 2.2 \pm 0.4$ eV in the case of Alq3 on pristine and $\Delta'_e = 2.1 \pm 0.4$ eV on MV0 precovered Ag(111). This is a rather small improvement, which is expected, since the work function reductions of MV0 and Alq3 on Ag(111) are comparable.

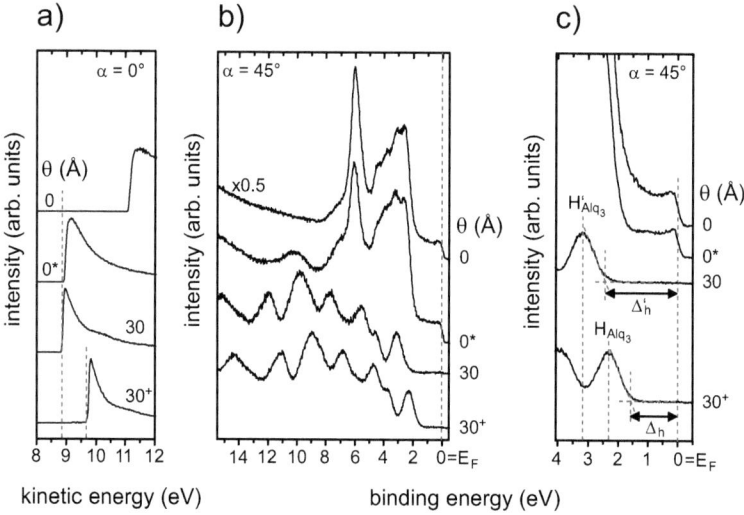

Figure 5.15.: Valence band and SECO spectra for the deposition of Alq3 on pristine ("30$^+$") and MV0 pre-covered ("30") Au(111). The thickness of the Alq3 film was in both cases 30 Å. The HOMO positions of Alq3 are labeled H'_{Alq_3} (MV0 pre-covered Au(111)) and H_{Alq_3} (pristine Au(111)) and the onsets of both peaks are Δ'_h and Δ_h respectively. The MV0 pre-covered spectrum is denoted by "0*" and the Fermi level by E_F.

In contrast, the work function reduction of MV0 on Au(111) is $\Delta\Phi_{Au,MV0} = -2.2$ eV (spectrum denoted "0*" in Fig. 5.15a) and only $\Delta\Phi_{Au,Alq_3} = -1.35$ eV for 30 Å Alq3 (spectrum denoted "30$^+$" in Fig. 5.15a). Since there is no change in the vacuum level during the deposition of 30 Å Alq3 on MV0 covered Au(111) (spectrum denoted "30" in Fig. 5.15a), the difference in the vacuum level almost directly translates into hole and electron injection barriers. This is shown in Fig. 5.15b and c, where for Alq3 on pristine Au(111) $\Delta_h = 1.60$ eV (HOMO H_{Alq_3} peak maximum = 2.35 eV BE) is found, while it is increased to $\Delta'_h = 2.40$ eV (HOMO H'_{Alq_3} peak maximum = 3.15 eV BE) in the case of MV0 covered Au(111). The resulting estimates for the electron injection barriers are $\Delta_e = 3.0 \pm 0.4$ eV in the case of Alq3 on pristine and $\Delta'_e = 2.2 \pm 0.4$ eV on MV0 precovered Au(111). Thus the insertion of

5.2. MV0 on coinage metals

an MV0 interlayer has effectively reduced the electron injection barrier by 0.8 eV, which is similar to the difference in work functions (0.85 eV) of both systems.

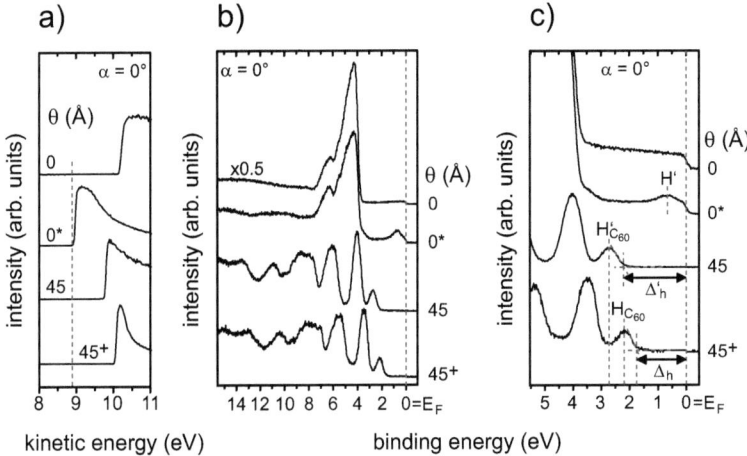

Figure 5.16.: Valence band and SECO spectra for the deposition of C_{60} on pristine ("45$^+$") and MV0 pre-covered ("45") Ag(111). The thickness of the C_{60} film was in both cases 45 Å. The HOMO positions of C_{60} are labeled $H'_{C_{60}}$ (MV0 pre-covered Ag(111)) and $H_{C_{60}}$ (pristine Ag(111)) and the onsets of both peaks are Δ'_h and Δ_h respectively. The MV0 pre-covered spectrum is denoted by "0*", the HOMO position of MV0 by H' and the Fermi level by E_F.

When C_{60} is used instead of Alq$_3$, the results are in principle similar, but an interface dipole builds up between MV0 and C_{60}. Due to the smaller transport gap (reports in the literature range from $E_t = 2.3 \pm 0.1$ eV [164] to $E_t = 2.6 \pm 0.1$ eV [165]) pinning of a C_{60} level just below the LUMO (P1 in Fig. 2.6) at the Fermi level occurs. Fig. 5.16a shows that the work function is initially reduced to $\Phi_{Ag,MV0} = 3.35$ eV by deposition of MV0 (spectrum labeled "0*"), but increases to 4.20 eV after 45 Å C_{60} deposition (spectrum labeled "45"). The work function of a similar amount (45 Å) of C_{60} deposited on pristine Ag(111) remains unchanged (spectrum labeled "45$^+$" in Fig. 5.16a), because the C_{60} molecules in direct contact with the substrate undergo a CT reaction, which increases the work function and exactly compensates for the "push-back" effect [166, 167]. The position of the C_{60} HOMO level (denoted H$_{C_{60}}$ in Fig. 5.16c) is 2.3 eV BE with resulting in an $\Delta_h = 1.70$ eV for C_{60} on pristine Ag(111). On MV0 precovered Ag(111) the HOMO position of C_{60} (denoted H'$_{C_{60}}$) is shifted to higher BE and located at 2.75 eV BE, while the hole injection barrier is now $\Delta'_h = 2.20$ eV. The resulting estimates for the electron injection barrier using the C_{60} transport gap are $\Delta_e = 0.6-$

Chapter 5. Results and Discussion

0.9 ± 0.1 eV in the case of C_{60} on pristine and Δ'_e = 0.1-0.4 ± 0.1 eV on MV0 precovered Ag(111). The increase in work function by +0.85 eV in the latter case is due to charge transfer from MV0 to C_{60}, which can also be regarded as Fermi level pinning of a C_{60} level just below the LUMO (P1), since it would be located below the Fermi level if the vacuum levels between MV0 and C_{60} would align.

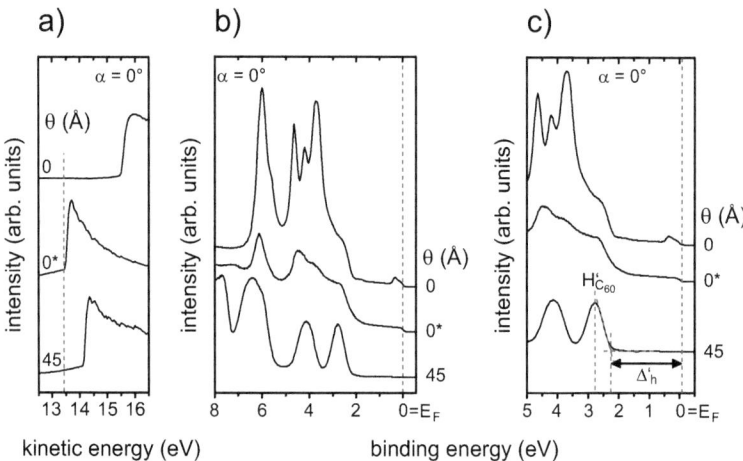

Figure 5.17.: Valence band and SECO spectra for the deposition of C_{60} on MV0 pre-covered Au(111). The thickness of the C_{60} film was 45 Å. The HOMO position of C_{60} is labeled $H'_{C_{60}}$ and its onset by Δ'_h. The MV0 pre-covered spectrum is denoted by "0*" and the Fermi level by E_F.

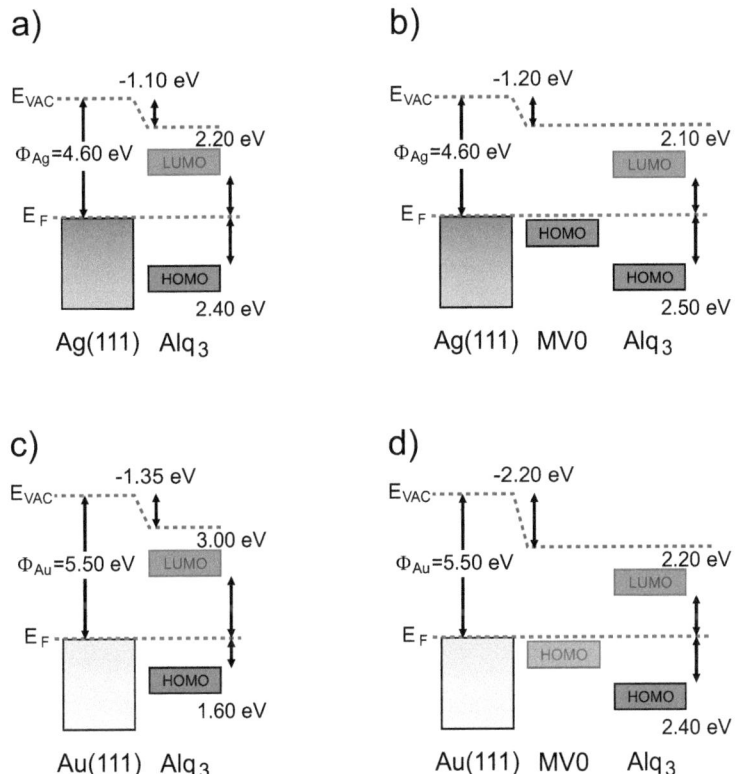

Figure 5.18.: Schematic energy level diagram for the deposition of Alq$_3$ on pristine and MV0 pre-covered Ag(111) (*upper part*) and Au(111) (*lower part*). The transport gap of Alq$_3$ was taken to $E_t = 4.6 \pm 0.4$ eV from Ref. [24]. E_F denotes the Fermi level and E_{VAC} the vacuum level. Partially transparent boxes show molecular orbitals that have not been resolved experimentally.

Chapter 5. Results and Discussion

The initial work function of MV0 precovered Au(111) of $\Delta_{Au,MV0} = 3.30$ eV (labeled "0*" in Fig. 5.17a) changes to 4.10 eV after the deposition of 45 Å C_{60} (labeled "45"). As shown in Fig. 5.17b and c, the HOMO position ($H'_{C_{60}}$) is located at 2.80 eV BE yielding a hole injection barrier of $\Delta'_h = 2.25$ eV. C_{60} deposited on pristine Au(111) [168] leads to a decrease in the work function of 0.60 eV, resulting in an absolute work function of 4.70 eV. The peak maximum of the HOMO is located at 2.20 eV BE leading to $\Delta_h = 1.60$ eV. The estimated electron injection barriers for the two systems are: $\Delta_e = 0.7\text{-}1.0 \pm 0.1$ eV (C_{60} on pristine Au(111)) and $\Delta'_e = 0.05\text{-}0.35 \pm 0.1$ eV (MV0 precovered Au(111)). Again CT from MV0 to C_{60} occurs, since an interface dipole of +0.8 eV is created similar to the case of C_{60}/MV0/Ag(111).

The energy level diagrams for the each four systems are summarized in Fig. 5.18 for Alq$_3$ and Fig. 5.19 for C_{60}.

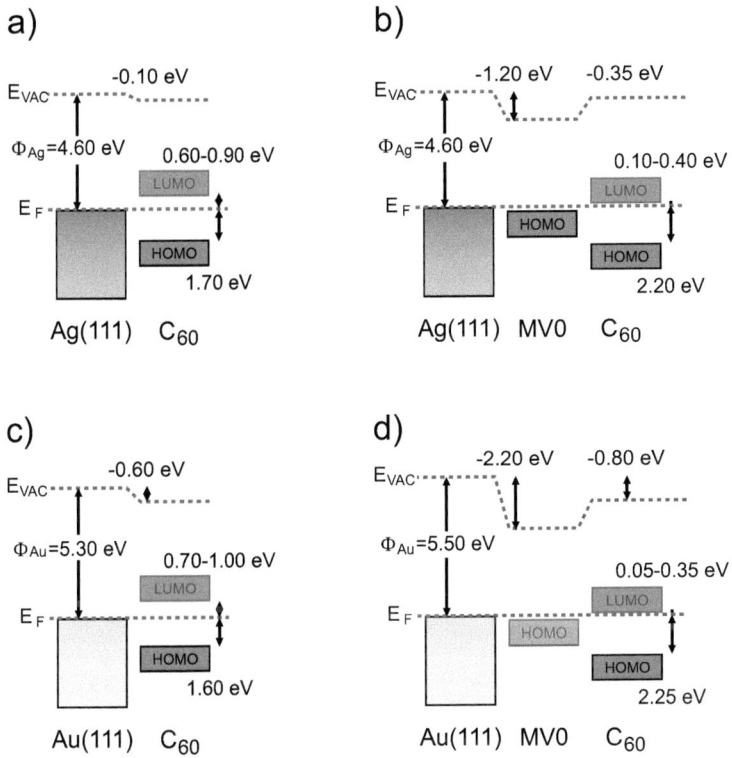

Figure 5.19.: Schematic energy level diagram for the deposition of C_{60} on pristine and MV0 pre-covered Ag(111) (*upper part*) and Au(111) (*lower part*). The transport gap of C_{60} was found in literature to range from $E_t = 2.3 \pm 0.1$ eV [164] to 2.6 ± 0.1 eV [165]. The values for C_{60} on pristine Au(111) are taken from Ref. [168]. E_F denotes the Fermi level and E_{VAC} the vacuum level. Partially transparent boxes show molecular orbitals that have not been resolved experimentally.

Chapter 5. Results and Discussion

5.2.7. Conclusions

In this chapter it has been shown that MV0 acts as a strong electron donor on Ag(111), Cu(111) and especially on Au(111). Deposition of MV0 in UHV resulted in a build-up of a molecular vapor phase, from which the adsorption on the substrates took place. After completion of a saturated monolayer, no change in the molecular emission features was observed in PES, which is an indication that multilayers are thermodynamically unstable. A strong work function reduction was observed for the saturated monolayer film on all three metal substrates, which was significantly larger than reported values for pure electron "push-back". This is direct evidence for a strong interaction between MV0 and the metal substrates including charge transfer from the molecule to the metal. Furthermore, the obtained work functions after saturated MV0 adsorption $\Phi_{mod} = 3.3$ eV are the same for all three coinage metals. This is ascribed to a pinning of the MV0 molecular HOMO at the Fermi level. This was also confirmed for MV0 adsorbed on Au(111) even though emission from the HOMO was not observed in the VB spectra here (possibly due to too low photoemission cross sections), but deeper lying MV0 molecular orbitals were found at the same binding energy position as for Ag(111) and Cu(111). The face-on orientation of the MV0 molecule on Ag(111) was confirmed independently by evaluation of the RAIR spectra and the angular dependence of the HOMO photoemission intensity. Since the PES fingerprint of MV0 on Cu(111) is almost identical to Ag(111) a similar orientation of the molecules in the saturated monolayer is inferred. For MV0 on gold an inclined geometry is proposed, which is supported by the observation of two peaks in the nitrogen core level spectra compared to only one in the case of silver and copper. Besides the information about the orientation of MV0, the RAIR spectra also showed that the vibrational fingerprint of MV0 adsorbed on Ag(111) is very similar to that reported for MV+1. This is an additional confirmation of the electron donation from MV0 to the metal substrate. Theoretical modeling was only done for MV0 adsorbed on Au(111), because here the strongest effect was experimentally observed. In the calculations a work function modification of -1.9 eV was found, which is in good agreement with the experimentally observed value of -2.2 eV. As origin the modeling also suggests a significant electron transfer from the molecule to the metal. The bonding pattern of MV0 was observed to change upon charge transfer from quinoid to benzoid, which stabilized the positive charge on the molecule and can be thought of a driving force for the electron transfer. The low work function surface of MV0 modified Au(111) and Ag(111) were further used to realize low electron injection barriers into subsequently deposited organic electron transport materials. The resulting decreases of Δ_e were 0.80 eV (0.1 eV) for Alq$_3$ and 0.65 eV (0.5 eV) for C$_{60}$ on Au(111) (Ag(111)). These experiments using MV0 therefore proved that the concept of injection barrier lowering also works at the cathode side of devices. However, the evaporation of MV0 close to room temperature preclude its use in actual devices, since the integration into a stable production process requires temperatures beyond 100 °C.

5.3. NMA as a high molecular weight donor for electron injection interlayers on metal electrodes

In Sec. 5.2 it was shown that thin layers of MV0 can reduce the work function of metal single crystals to values of $\Phi_{mod} = 3.3$ eV, which is below the value of pristine magnesium ($\Phi_{Mg} = 3.7$ eV [169]). Using these modified electrodes, Δ_e into Alq$_3$ and C$_{60}$ could be significantly reduced. Consequently, this presents a promising way for using high Φ metal cathodes that are easy to handle but still provide efficient electron injection when an appropriate donor layer is applied. To solve the issue of low evaporation temperatures of MV0, molecules with a higher molecular weight and rather non-planar gas phase structures could be used. NMA is one candidate in this context, since the donor molecule is rather large and has a pronounced 3D structure in the gas phase. In the following, its molecular properties at interfaces with metal substrates and its potential to lower Δ_e into subsequently deposited materials are explored. This section was published in Ref. [98].

5.3.1. Introduction

In this study, NMA was chosen as molecular donor because its chemical structure exhibits pronounced similarities to the strongly Φ-reducing MV0 [150] and the screening has shown a strong work function modification potential; both molecules are quinoidal in the neutral state and become aromatic upon oxidation, which stabilizes the charge, and both molecules contain two azine rings. A distinct advantage of NMA over MV0 is its higher molecular weight (MV0: 186 g/mol, NMA: 412 g/mol). The calculated vertical $IE_{vert,NMA}$ of NMA is $IE = 5.45$ eV, which is somewhat larger than the calculated $IE_{vert,MV0} = 4.65$ eV of MV0. Nevertheless, it is still lower than the IE of tetrathiafulvalene (6.3 eV calculated using the same methodology [154, 170], which is well known for its donor properties [171, 172]). Therefore, the choice of NMA as potential donor material for lowering the work function of metal surfaces is well justified. In the following, the characterization of NMA by several experimental techniques is described, and its suitability and performance as molecular donor at organic electronic device interfaces is assessed.

5.3.2. Photoelectron spectroscopy at interfaces to metals

Sequential deposition of NMA onto Au(111) (Fig. 5.20a and b) resulted in a fast attenuation of Au photoemission features and an emergence of new molecular derived features in the valence energy region. The Au surface state (centered at 0.25 eV BE, denoted with S in

Chapter 5. Results and Discussion

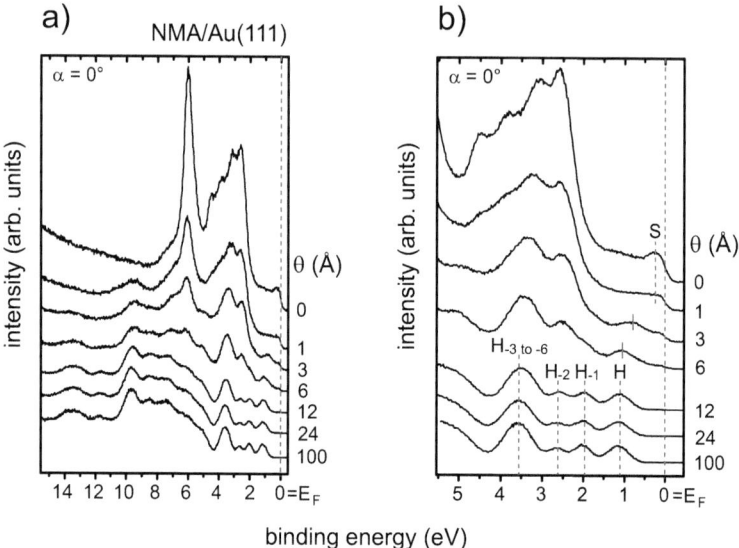

Figure 5.20.: a) Valence band spectra for the sequential deposition of NMA on Au(111). b) Zoom into the region close to the Fermi-edge. Zero coverage denotes the pristine Ag(111) spectrum and E_F the position of the Fermi-edge.

Fig. 5.20b), which is characteristic of a clean surface, vanished with nominally 1 Å NMA coverage. Already at 12 Å NMA film thickness, the Au 5d and 6s valence features (from 7 to 2 eV BE and up to the Fermi level) are no longer visible in the spectrum.

Similar observations are made for NMA deposited onto Ag(111) and Cu(111), as shown in Figs. 5.21 and 5.22. This suggests layer-by-layer growth in the initial phases of thin film growth on all three substrates. The peak positions of the six highest occupied molecular orbital levels of NMA on the three metal substrates are summarized in Tab. 5.1. In essence, the peak positions rigidly shift towards higher BE with increasing film thickness until a coverage of 12 Å NMA is reached on all three substrates. For higher coverages the positions remain unchanged. No difference in the spectra were observed when changing the take-off angle to $\alpha = 45°$.

The molecular structure and DOS of NMA in the gas phase was calculated using DFT methodology and the Gaussian software as outlined in Sec. 4.5. These calculations were done by Oliver T. Hofmann (TU Graz, Austria) and Georg Heimel (HU Berlin, Germany). The neutral molecule is found to be severely strained, as the quinoidal bridge tries to enforce

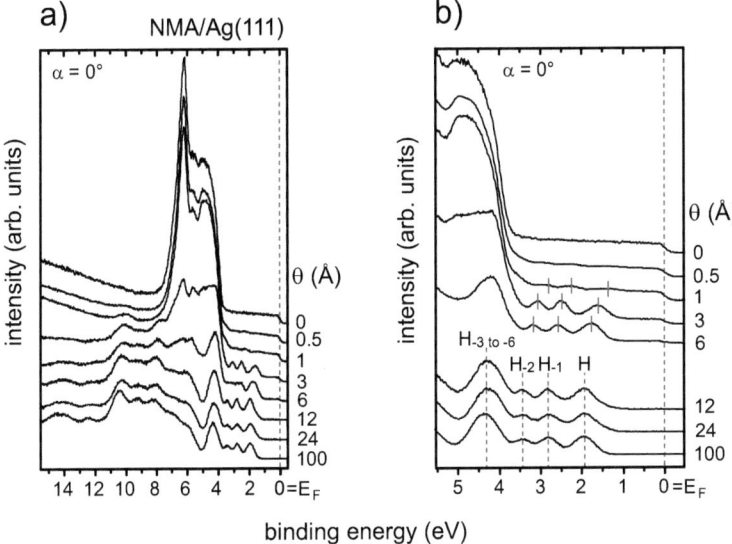

Figure 5.21.: a) Valence band spectra for the sequential deposition of NMA on Ag(111). b) Zoom into the region close to the Fermi-edge. Zero coverage denotes the pristine Ag(111) spectrum and E_F the position of the Fermi-edge.

a close to planar backbone, which results in strong steric repulsions between the hydrogen atoms on the acridanic ring and the bridging ethenyl group. This results in heavily distorted geometries, where the acridanic rings are not planar, but rather adopt a boat-like structure and are twisted with respect to each other. Two (local) minimum configurations have thus been investigated, which differ only in the bending of the rings. In the energetically more favorable geometry, both "boats" point into the same direction (Conf. A in the upper panel in Fig. A.4 b in the appendix), while the other geometry, being energetically higher by about 6.8 kJmol^{-1}, has the two "boats" pointing in opposite directions (Conf. B in the lower panel in Fig. A.4 b in the appendix). In Fig. 5.23, the theoretical DOS of neutral NMA (without taking into account photoemission cross sections) is compared to the UPS spectra of nominally 12 Å thick NMA films on the three metal substrates. For better comparison with the experiment a Gaussian broadening (FWHM = 0.5 eV) was applied. All spectra were aligned with their HOMO to the HOMO position of NMA on Au(111). The respective shifts were 0.83 eV for NMA on Ag(111) (Fig. 5.23b), 0.65 eV in the case of Cu(111) (Fig. 5.23c) and 1.30 eV for the calculated DOS (Fig. 5.23d). The fact that already at 12 Å film thickness the UPS spectra

Chapter 5. Results and Discussion

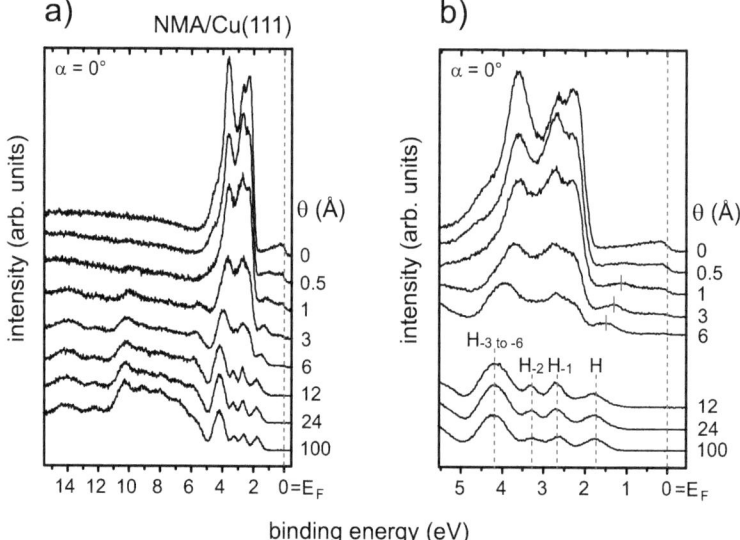

Figure 5.22.: a) Valence band spectra for the sequential deposition of NMA on Cu(111). b) Zoom into the region close to the Fermi-edge. Zero coverage denotes the pristine Cu(111) spectrum and E_F the position of the Fermi-edge.

are governed solely by molecular features indicates that molecules in the multilayer are not influenced by the substrate, which is in part related to the distorted molecular structure discussed above. This is similar to the case of rubrene on Au(111) [173] and might lead in the present case to Frank-van der Merwe (i.e, layer by layer) growth.

	NMA film thickness	HOMO (eV)	HOMO-1 (eV)	HOMO-2 (eV)	HOMO-3 to HOMO-6 (eV)
Au(111)	3 Å	0.85	-	-	-
	12 Å	1.15	2.00	2.60	3.50
Ag(111)	3 Å	1.60	2.50	3.10	-
	12 Å	1.95	2.80	3.45	4.30
Cu(111)	3 Å	1.35	-	-	-
	12 Å	1.80	2.70	3.30	4.20

Table 5.1.: Peak positions of the six highest molecular orbitals of NMA adsorbed on the (111) surfaces of Au, Ag, and Cu single crystals. Peak positions are given for 3 and 12 Å film thickness. For the thin films the positions of the deeper lying peaks cannot always be determined because of the overlap with metal d bands.

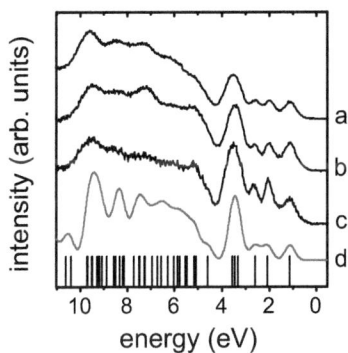

Figure 5.23.: Comparison of the valence band spectra of a 12 Å thick film of NMA on a) Au(111), b) Ag(111), and c) Cu(111). The DFT-calculated spectrum is plotted in red (d). The vertical black bars correspond to the calculated orbital energies. All spectra have been aligned with their HOMO to the HOMO position of NMA on gold. For the values of the necessary shifts see text. All spectra are normalized to their maximum intensity value, none of the spectra has been stretched or compressed on the energy scale.

Chapter 5. Results and Discussion

The evolution of the sample work function for increasing NMA coverage as measured from the SECO is shown in Fig. 5.24a for all three metal single crystal substrates. In all cases, Φ is significantly decreased upon NMA deposition. At 12 Å NMA coverage the modified Φ of Au is 4.10 eV, which is 1.40 eV lower than that of pristine Au(111) (5.50 eV). Literature values for metal surface electron "push-back" [38, 46] induced change for Au range from −0.53 eV (Xe [174]) to −0.7 eV (long-chain alkanes like TTC [43]) or even −1.3 eV (α-NPD) [44, 9]). On silver, Φ changes from 4.60 (pristine Ag(111)) to 3.60 eV at 12 Å NMA film thickness. Similarly, Φ changes from 4.90 eV (pristine Cu(111)) to 3.75 eV (12 Å NMA) on copper. In both cases, the change is ≈0.3 eV larger than reported values (Ag(111) ≈0.7 eV, Cu(111) ≈0.9 eV [175, 176]) for weakly interacting molecules on silver and copper. Thus, a (weak) charge transfer that decreases Φ below the reported values for the "push-back" effect can be assumed in all three cases, despite the fact that the VB spectra show no direct evidence for a new density of states (intra-gap states) or pronounced differential shifts of the occupied molecular orbitals [11]. Related observations have been reported for the acceptor octafluoroanthraquinone, which increased Φ of Ag(111), but does not show clear new intra-gap states resulting from the charge transfer in the valence band spectra [177]. In this context, one has to keep in mind that due to a hybridization of molecular and metal states the effective occupation of certain molecular orbitals can be changed by charge transfer processes even if the associated DOS peaks are far below (or in the case of acceptors above) the Fermi level [16].

The molecular IE of NMA on gold is ca. 4.65 eV for coverages of 12 Å and above. The bulk IE value is ca. 0.30 eV higher for NMA films on silver and copper (both ca. 4.95 eV) substrates. The reason for this observation may be a different orientation of the molecules on gold compared to that on silver and copper substrates, as the IE of organic molecules is highly orientation dependent [30, 31]. A strong attenuation of the substrate's photoemission features is not only observed in the valence region by UPS, but also in the corresponding core level spectra by XPS. Due to the higher information depth at the electron energies of XPS, photoelectrons from the substrate core levels are still detected even at 24 Å NMA film thicknesses. They become fully attenuated between 24 and 100 Å coverage. The peak areas of the Au 4f, Ag 3d, and Cu 3p core levels are plotted in Fig. 5.24b, c, and d as function of the nominal film thickness. They have been normalized with respect to the investigated volume as explained in Sec. 3.1.6. The resulting data points have all been fitted using a *single* exponential decay function of the form:

$$\frac{I(d)}{I(0)} = k \cdot \exp^{\frac{-d}{\lambda}} \quad (5.1)$$

where k is a constant, d the nominal film thickness, and λ the elastic mean-free electron path. This yields $\lambda = 5.5\pm1$ Å (Cu), $\lambda = 4.5\pm1$ Å (Au) and $\lambda = 3.5\pm1$ Å (Ag). These values

5.3. NMA on coinage metals

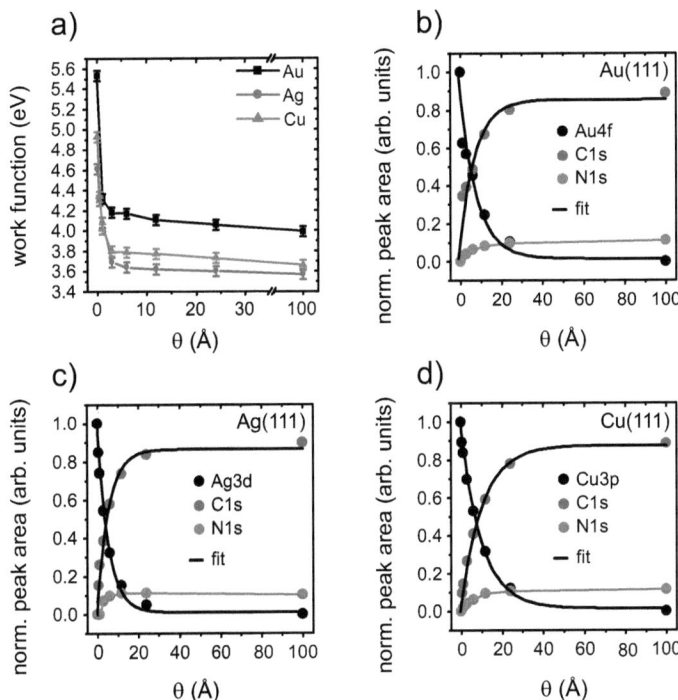

Figure 5.24.: a) Absolute work function of the NMA modified metal single crystals as function of NMA film thickness θ. b), c), d) Normalized integrated peak area of the C1s (red dots), N1s (green dots), and the substrate's core levels (black dots). They have been normalized with respect to the investigated volume as explained in Sec. 3.1.6. The black lines represent fits to the data points using a single exponential decay function.

deviate slightly (by 5-7 Å) from simulated values found in literature [80, 178]. Additionally, the C1s and N1s core levels have been fitted on all three substrates as shown in Figs. 5.25 (C1s) and 5.26 (N1s) for 3 and 12 Å coverage. For the full core level spectra see Figs. A.5, A.6, and A.7 in the appendix.

From the spectral deconvolution two contributions in the C1s core level spectra can be distinguished for all coverages. The peak denoted C2 originates from carbon atoms bonded to nitrogen as marked by the red dots in Fig. 5.27. All other carbons contribute to the peak labeled C1 (marked by black dots in Fig. 5.27). In the spectra of the N1s core levels, only one

Chapter 5. Results and Discussion

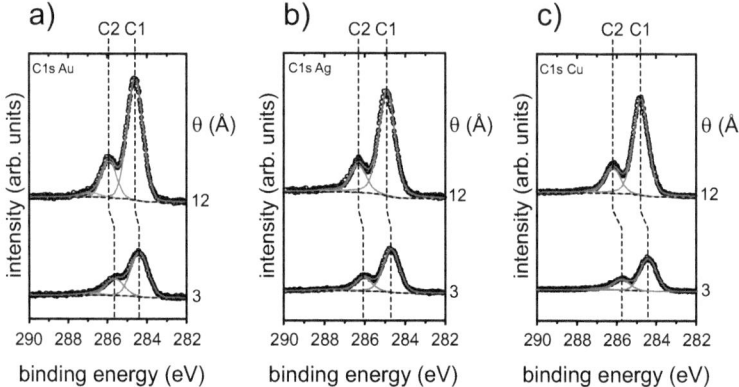

Figure 5.25.: Carbon 1s core level spectra for 3 and 12 Å of NMA on a) Au(111), b) Ag(111), and c) Cu(111). Spectral deconvolution has been done using mixed Gaussian and Lorentzian peaks and Shirley backgrounds. Two contributions (C1 and C2) are observed in the spectra, which shift to higher BE with increasing coverage as indicated by the dashed black lines.

single peak with a constant FWHM is observed. Thus, the nitrogen atoms in the molecule have always equivalent chemical environments, even close to the surface in the sub-monolayer regime (0.5, 1, 3, 6 Å). The integrated and normalized peak areas of the carbon and nitrogen 1s peaks are plotted in the graphs in Fig. 5.24b, c, and d. The ratio between C1s and N1s is close to the ideal stoichiometric ratio of 15:1 (C:N) in all three cases (max. deviation 20%). The C1s data points have also been fitted using a *single* exponential growth function as shown by the black line in the graphs. The exponential decay of the substrate's core levels and the growth of the C1s and N1s core levels indicates homogeneous film formation on all three substrates [179]. Consistent with the above discussed UPS and substrate core level results, it is proposed that the growth mode on all three substrates is (at least to a good approximation) layer-by-layer.

The coverage dependent evolution of the integrated peak areas for the two C1s contributions C1 and C2 are plotted in Fig. 5.28. On Ag(111) and Cu(111) the exponential growth is similar for both, C1 and C2, and their ratio is 3.5:1 (C1:C2), which is close to the expected stochiometric ratio of 4:1. In contrast, a larger C2 peak is observed for the lowest coverage (1 Å) on Au(111), indicated by the blue arrow in Fig. 5.28, where a ratio of 1.4:1 (C1:C2) is found. At 3 Å the ratio is 2:1 and at 6 Å a ratio of 3.6:1, which remains more or less constant for higher coverages, similar to the case on Ag(111) and Cu(111). This indicates, that some carbon atoms have a different chemical environment for low NMA coverages on

5.3. NMA on coinage metals

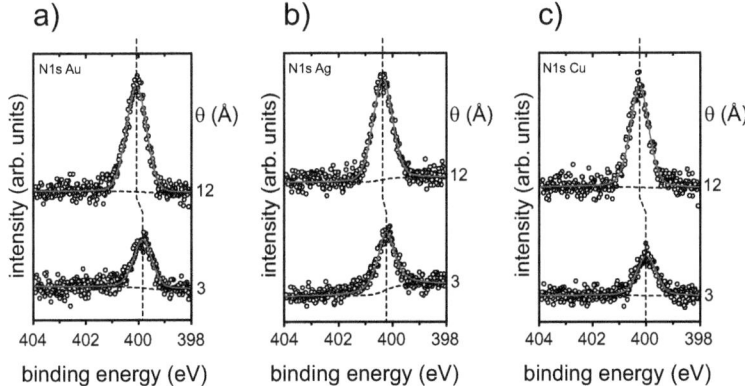

Figure 5.26.: Nitrogen 1s core level spectra for 3 and 12 Å of NMA on a) Au(111), b) Ag(111), and c) Cu(111). Spectral deconvolution has been done using mixed Gaussian and Lorentzian peaks and Shirley backgrounds. A single peak is observed in the spectra, which shifts to higher BE with increasing coverage as indicated by the dashed black lines.

Au(111) pointing towards stronger interaction with the substrate. This possibly involves charge rearrangements and results in a chemically shifted core level. In all three experimental sets, the C1s and N1s core level spectra have been further analyzed to compare NMA coverage dependent core level shifts to shifts observed in the valence band for sub- and monolayer coverages (3 and 12 Å). The results are summarized in Tab. 5.2 and also indicated by the dashed red lines in the spectra plotted in Figs. 5.25 (C1s) and 5.26 (N1s). All core and valence levels are observed to shift rigidly, except for the N1s core level on silver and copper. The reduction by 0.1 eV to 0.2 eV may actually be an indication for a direct interaction between the molecules and the silver and copper surfaces, involving possibly the lone pairs of

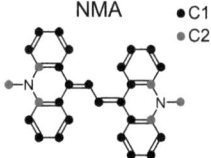

Figure 5.27.: Chemical structure of NMA indicating the two different chemical environments of the carbon peaks C1 and C2 in Fig. 5.25: carbon atoms bonded to nitrogen (C2, red dots) and all others (C1, black dots).

101

Chapter 5. Results and Discussion

molecular orbital(s)	shift on Au(111) (eV)	shift on Ag(111) (eV)	shift on Cu(111) (eV)
HOMO	0.30	0.35	0.45
HOMO-1	-	0.30	-
HOMO-2	-	0.35	-
C1s (C1)	0.25	0.20	0.40
C1s (C2)	0.30	0.25	0.40
N1s	0.25	**0.15**	**0.15**

Table 5.2.: Comparison of UPS and XPS shifts when going from 3 to 12 Å NMA coverage; top three lines: UPS measured shifts of different NMA orbitals on Au(111), Ag(111), and Cu(111). Where no values are given, the peaks cannot be clearly distinguished from the substrate d bands. Bottom three lines: XPS shifts of the low (C1) and high (C2) binding energy C1s peaks and the N1s peak.

the nitrogen atoms.

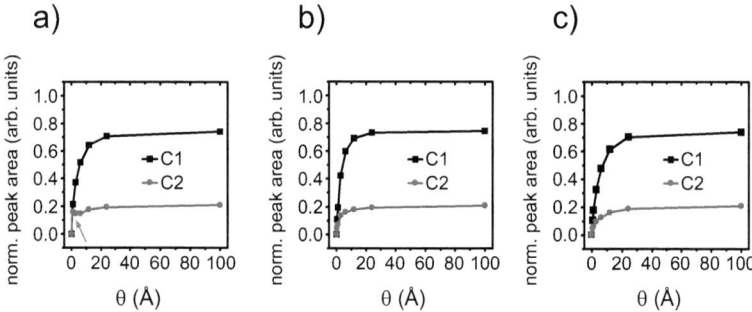

Figure 5.28.: Coverage dependent evolution of the integrated peak areas for the two C1s contributions C1 and C2. The blue arrow denotes an increased C2 integrated peak area, which results in an different ratio between C1 and C2.

5.3.3. Electron injection barrier tuning with NMA interlayers

To demonstrate that the Φ decrease due to NMA deposition onto the metal surface results indeed in the expected lowering of Δ_e into subsequently deposited organic films, Alq$_3$ was deposited onto the NMA-modified Au(111). Then VB spectra were measured. The experiments were conducted with three different NMA pre-coverages on Au(111) (3 Å, 6 Å, and 12 Å) in order to analyze the relationship between the modified substrate Φ_{mod} and Δ_e. As reference system, the uppermost trace in Fig. 5.29a displays the valence band spectrum of 30 Å Alq$_3$ on pristine Au(111). The HOMO onset of Alq$_3$ is located at 1.60 eV BE, which

5.3. NMA on coinage metals

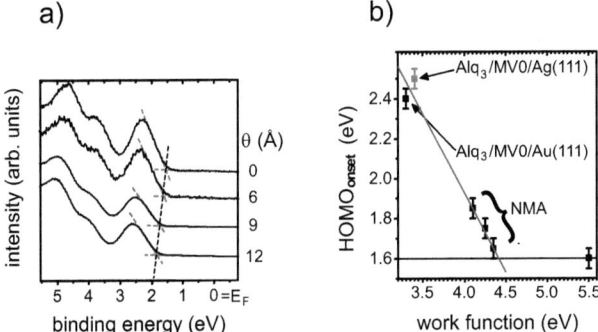

Figure 5.29.: a) Valence band spectra of 30 Å Alq3 on pristine (0 Å) and 6, 9, and 12 Å NMA precovered Au(111). E_F denotes the Fermi level. b) Alq3 HOMO onset plotted versus the work function of the modified/pristine Au(111) substrate. The red line is a linear fit (slope=−0.75) to the three data points for NMA precovered Au(111). For comparison, also the data points for Alq3 on MV0 precovered Au and Ag from Fig. 5.18 are included.

translates into an Δ_e of 3.00 ± 0.40 eV, assuming the reported transport gap of $E_t = 4.60$ ± 0.40 eV for Alq3 [24]. Pre-covering of the Au substrate with (sub-) monolayers of NMA before the deposition of 30 Å Alq3, results in a rigid shift of all Alq3 molecular levels (i.e., low energy HOMO onset and peak maximum) to higher BE, the magnitude of the shift depending on the pre-coverage amount. This is also shown in Fig. 5.29a, where the observed values for the HOMO onsets of Alq3 are 1.65 eV BE (6 Å NMA, $\Delta_e = 2.95 ± 0.40$ eV), 1.75 eV BE (9 Å NMA, $\Delta_e = 2.85 ± 0.40$ eV), and 1.85 eV BE (12 Å NMA, $\Delta_e = 2.75 ± 0.40$ eV). Fig. 5.29b shows the Alq3 HOMO onset dependence on the substrate. The three data points representing NMA pre-covered substrates can be fitted by a linear function with a slope of $S = -0.75$, as shown by the red line. This deviation from the Schottky-Mott limit ($S = -1$, [47]) hints that Alq3 deposition does still influence the interface dipole of the donor modified Au(111) substrate (i.e., depolarization might take place, or the conformation of adsorbed NMA is changed due to the overlayer). Interestingly, the data points of Alq3 on MV0 modified Au(111) and Ag(111) (as shown in Sec. 5.2.6) also fit to this line. There, the Au(111) and Ag(111) substrates were modified using MV0, which resulted in a $\Phi_{Au,MV0}$ of 3.3 eV. The HOMO onset of subsequently deposited Alq3 (30 Å) was measured to be 2.4 eV BE (Au) and 2.5 eV BE (Ag). This indicates that the slope of −0.75 is not a parameter that depends on the molecule used for the work function/interface dipole modification, but may be indeed related to electronic properties of Alq3 itself. Moreover, the offset between 5.5 eV

103

Chapter 5. Results and Discussion

and 4.5 eV shows a region where not tuning of Δ_e is possible due to the intrinsic "push-back" effect.

5.3.4. Conclusions

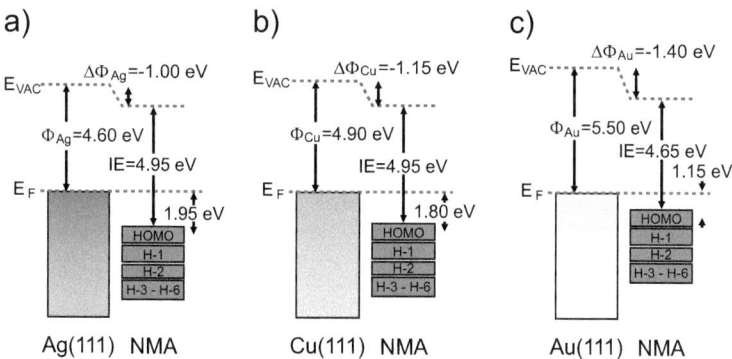

Figure 5.30.: Energy level diagram of a 12 Å NMA film on a) Au(111), b) Ag(111), and Cu(111). E_F denotes the Fermi level and E_{VAC} the vacuum level.

The adsorption and electronic structure of NMA films on the (111) faces of silver, copper, and gold single crystals have been investigated by PES. The work function decrease observed in the SECO spectra was beyond what has previously been reported for pure electron "push-back", however, not as strong as found for MV0. Nevertheless, the findings indicate a (weak) charge transfer between the electron donor NMA and the respective metals. This is further supported by quantitative analysis of th C1s and N1s core level spectra: On Ag(111) and Cu(111) the spectra are very similar and in both cases a larger shift of the N1s peak to higher BE compared to the other molecular levels is found. On Au(111) the ratio between both C1s peaks for sub-monolayer films is significantly different compared to the multilayers. This points to chemical interaction, which energetically shifts the C1 carbon atoms to higher binding energy, which changes the respective peak areas. In all three cases the observations support an interaction beyond pure physisorption. The different behavior of NMA on silver and copper compared to gold is also reflected in the IE, which is equal for Ag and Cu but 0.3 eV higher on Au(111). The valence band spectra showed that the position of the NMA molecular orbitals remained constant for nominal coverages above 12 Å, which is thus regarded as a closed monolayer. The results of the UPS experiments are summarized in the energy level diagrams in Fig. 5.30. Furthermore, the fast attenuation of the substrate features in the VB is also seen in the attenuation of the substrate's core levels. The integrated peak area shows a single exponential decrease, which is indicative of layer-by-layer growth. Finally, the low work function of NMA modified Au(111) were used successfully to decrease Δ_e into subsequently deposited Alq$_3$ films. The lowering of the barrier was however restricted to a maximum of 0.25 eV, which is far below the value of MV0 (−0.80 eV). Judging from injection barrier tuning and the work function decrease, only weak donor properties are attributed to NMA. Nevertheless, its thermal properties are superior to MV0, since the evaporation temperature is far above 100 °C. Molecular structures similar to NMA are thus promising for new electron donor materials.

5.4. The model acceptor F4-TCNQ adsorbed on Ag(111)

5.4.1. Introduction

As already discussed in Sec. 1, F4-TCNQ is a strong electron accepting molecule, which has been reported to increase the work function of polycrystalline gold ($\Delta\Phi_{Au,F4-TCNQ}$ = +0.35 eV) and Cu(111) ($\Delta\Phi_{Cu,F4-TCNQ}$ = +0.60 eV) [9, 16]. The origin for the work function increase has been ascribed to a combination of adsorption-induced geometric distortion of the molecules, charge transfer from the metal to the molecules, and molecule-to-metal back transfer. In this section the interaction between F4-TCNQ and Ag(111) is analyzed using UPS, RAIRS and theoretical modeling. Corroborating the findings on gold and copper, an increase of the work function on Ag(111) of $\Delta\Phi_{Ag,F4-TCNQ}$ = +0.65 eV is found. The valence band spectra clearly show two additional states, that are assigned to metal-molecule hybrid states derived from the former HOMO and LUMO of the neutral molecule, indicating significant net CT from the metal to the molecule. This is confirmed by DFT modeling, where a net electron transfer of 0.56 electrons is found. The RAIRS experiments further support the negative charging of the molecule on the surface by monitoring the position of the symmetric CN stretching vibration for mono- and multilayers. Furthermore, it is found that the molecules adopt a face-on conformation for (sub-)monolayer coverage on Ag(111) similar to the other coinage metal substrates. In the multilayers, the orientation changes and the long and short molecular axis become tilted with respect to the plane of the surface. Part of this work is published in Ref. [180].

5.4.2. Valence electronic structure

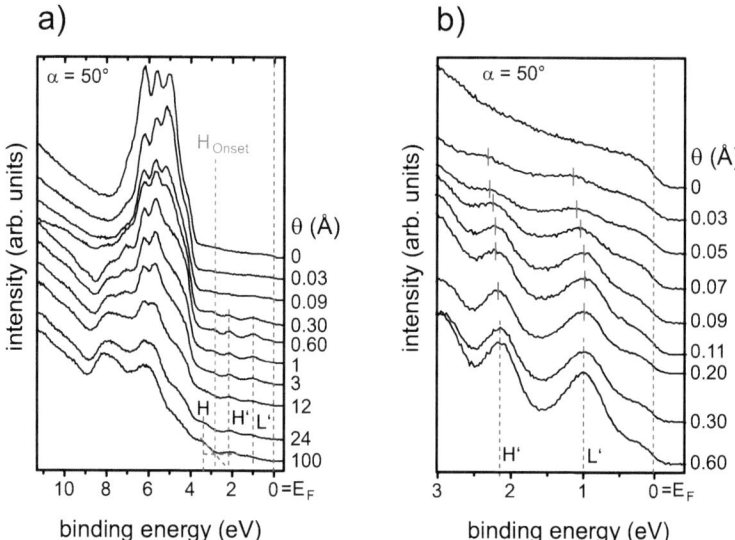

Figure 5.31.: Valence band spectra for the deposition of F4-TCNQ on Ag(111) as a function of nominal coverage θ. a) Full valence band and b) zoom into the region close to the Fermi level. H' and L' denote metal-molecule hybrid states derived from the HOMO (H') and LUMO (L') of neutral F4-TCNQ. H denotes the HOMO of the bulk molecular film, H_{Onset} its emission onset, and E_F the Fermi level.

F4-TCNQ has been deposited sequentially on Ag(111) and VB and SECO spectra have been recorded. The VB spectra are plotted in Fig. 5.31a for various coverages. With increasing nominal coverage of F4-TCNQ, the Ag photoemission features (4d and 5s bands) decrease in intensity and molecule derived features appear in the spectra. However, even at 100 Å the Fermi-edge is not fully attenuated, which suggests pronounced island growth. Fig. 5.31b shows a zoom into the low BE region of the VB for nominal coverages up to 0.6 Å. Here, two features can be clearly observed, which are molecule-metal hybrid states derived from the former LUMO (L') at 1.15 eV BE and HOMO (H') at 2.30 eV BE of the neutral molecule (both peak positions are obtained from the lowest coverage spectrum of 0.03 Å). The assignment is made in agreement with DFT calculations done using the VASP code and DFT methodology, as described in Sec. 4.5 (done by Gerold M. Rangger, TU Graz, Austria). Furthermore,

5.4. The model acceptor F4-TCNQ adsorbed on Ag(111)

similar assignments have been made for the adsorption of F4-TCNQ on Au and Cu(111) [9, 16]. Both features shift to lower BE with increasing F4-TCNQ coverage to reach a final position of 1.00 eV BE (L') and 2.15 eV BE (H') at 0.3 Å. The reason for this shift lies in different occupations of the L' state, which is increased by about 10 % for the submonolayer case (corresponding to 0.03 and 0.05 Å experimental coverage) compared to the monolayer case (0.3 to 0.6 Å) as obtained by the DFT modeling. The resulting theoretical shift of 0.2 eV to lower BE agrees very well with the experimentally found value of 0.15 eV as shown in Fig. 5.32a. Upon increasing the nominal coverage of F4-TCNQ beyond 0.6 Å, both features (H' and L') become slowly attenuated, which confirms their confinement to the interface. Their slow attenuation (they can even be observed in the 100 Å spectrum) supports island growth. For F4-TCNQ coverages above 3 Å features of the neutral bulk film are observed. The HOMO peak maximum is found at $\approx$ 3.4 eV BE as shoulder (denoted H in Fig. 5.31a) with its HOMO onset (H_{Onset}) at $\approx$ 2.9 eV BE. The molecule-metal hybrid state L' appears in the spectra, because charges are transferred from the metal to the molecule thus filling up a state derived from an initially unoccupied state of the neutral molecule. As a result the molecule becomes negatively charged and a dipole is created across the interface.

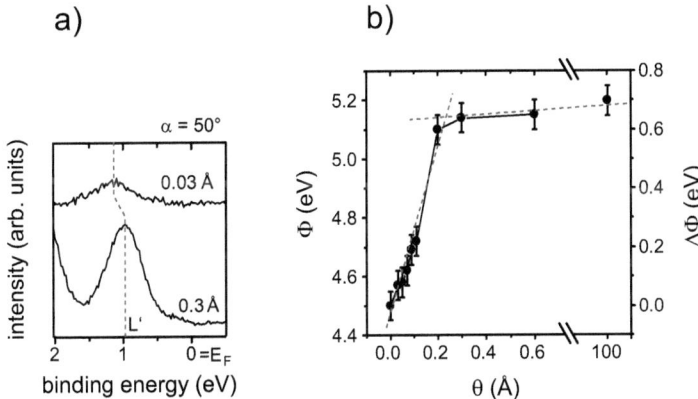

Figure 5.32.: a) detail of the valence band spectra shown in Fig. 5.31 showing the metal-molecule hybrid state derived from the LUMO of neutral F4-TCNQ (L') for very low coverages (0.03 Å) and coverage of about one monolayer (0.3 Å). The shift between both peak maxima is 0.15 eV. Note that the underlying photoemission features of the Ag(111) substrate have been subtracted in this plot. b) absolute work function Φ and relative work function change $\Delta\Phi$ (with respect to the pristine Ag(111) substrate) for different coverages of F4-TCNQ on Ag(111).

Chapter 5. Results and Discussion

This increases the work function of the system (as shown in Fig. 5.32b) from 4.5 eV (pristine Ag(111)) almost linearly to a value of $\Phi_{Ag,F4-TCNQ} = 5.15$ eV at 0.3 Å nominal coverage. At higher coverages the work function passes into a saturation regime, where it remains almost constant. This is expected, since a dipole across the interface can only be created by molecules in the very first monolayer, which are in direct contact with the substrate. Based on this finding, a nominal coverage of around 0.3 to 0.6 Å can be associated with an almost full monolayer (this agrees also well with the maximum in intensity of the L' and H' states, which occurs around 0.6 Å). The slight increase in work function for higher coverages to 5.20 eV at 100 Å occurs because remaining free spots on the surface (between the islands) are only slowly filled. The IE of neutral F4-TCNQ molecules in the bulk is derived as $IE = 8.1$ eV, which is in good agreement with the literature value of 8.34 eV [14].

5.4.3. DFT results

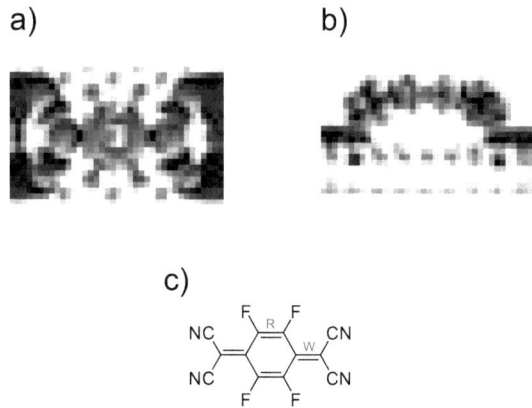

Figure 5.33.: *Upper part:* Top (a) and side view (b) of the three-dimensional charge-density rearrangements for F4-TCNQ adsorption on Ag(111). Electrons flow from the dark gray to the light gray areas. Only the top two metal rows are shown. *Lower part:* Chemical structure of F4-TCNQ (c). R and W indicate C=C double bonds of the Ring (R) and Wing (W) of both molecules.

The following theoretical results have been obtained by Gerold M. Rangger (TU Graz, Austria).

F4-TCNQ has a planar structure in the gas phase and belongs to the point group D_{2h}. The theoretical DFT modeling (using the VASP code) shows that upon adsorption on Ag(111) this situation changes dramatically: F4-TCNQ adopts a bent geometry with the nitrogen atoms

5.4. The model acceptor F4-TCNQ adsorbed on Ag(111)

1.23 Å closer to the top metal layer than the π backbone as shown in Fig. 5.33. This indicates a strong attractive interaction between the CN groups and the metal surface. The central carbon ring of the molecule is calculated to be 3.61 Å above the top Ag layer. Similar results have been obtained for F4-TCNQ adsorbed on Au(111) and Cu(111) surfaces and in case of the latter, the theoretical modeling has been confirmed experimentally by X-ray standing-wave experiments [16]. The point group of the adsorbed molecule is consequently reduced to (at least) C_{2v}. In the theoretical calculations, the work function change can be partitioned into two contributions: one arises from the charge rearrangements (fore- and back-donation, as indicated in Fig. 5.33 [16, 180]) and accounts in the present case for $\Delta_{CT} = +1.70$ eV (net charge transfer 0.56 electrons[3]). This would result in a strong work function increase. However, the second contribution comes from the bent conformation of the molecule (the geometric distortion induced by the adsorption), which creates a dipole that counteracts Δ_{CT}. It amounts for $\Delta_{Mol} = -0.85$ eV in the present case. Both contributions summed up lead to the total theoretically obtained work function change for the densely packed monolayer of $\Delta\Phi_{theo} = +0.85$ eV. The measured maximum work function increase of about 0.65 eV for F4-TCNQ on Ag(111) agrees very well with this calculated value, especially considering that the calculations assume a completely covered and perfectly ordered layer.

5.4.4. RAIRS results

The RAIR spectra recorded for the in-situ deposition of F4-TCNQ on Ag(111) are shown in Fig. 5.34. For sub-monolayer coverages (up to the red spectrum, which corresponds to nominally 0.6 Å) only five vibrational modes are observed at 1423, 1488, 1648, 2098, and 2170 cm^{-1}. The low number of observed vibrational modes is a first hint, that F4-TCNQ adopts a rather face-on adsorption geometry, since all modes above 800 cm^{-1} are in-plane of the molecule and thus suppressed in a face-on geometry (due to the surface selection rule). In the face-on geometry only totally symmetric modes are allowed and will be observed if they create a *significant* dynamical dipole normal to the surface as discussed in Sec. 3.2.2. Furthermore, a face-on geometry as shown in Fig. 5.33 has been observed experimentally on Au(111) and Cu(111) [16, 181]. It is also supported by the DFT modeling discussed in the previous section. The calculated charge-density rearrangements (Fig. 5.33a and b) show the areas of the F4-TCNQ molecule, which are especially influenced by the charge transfer with the substrate. These are the CN groups and the central carbon ring system. Thus, vibrational modes, which lead to displacements of atoms in these areas of the molecule are likely to create large dynamical dipoles perpendicular to the surface [182, 183]. In the

---

[3]The charge transfer into the hybrid state derived from the former LUMO of F4-TCNQ is approximately 1.8 electrons, which results in an almost completely filled molecular orbital (filling of 89 %). However, also back-donation from deeper lying molecular orbitals occurs, and as a consequence a net metal to molecule transfer of 0.56 electrons is obtained.

Chapter 5. Results and Discussion

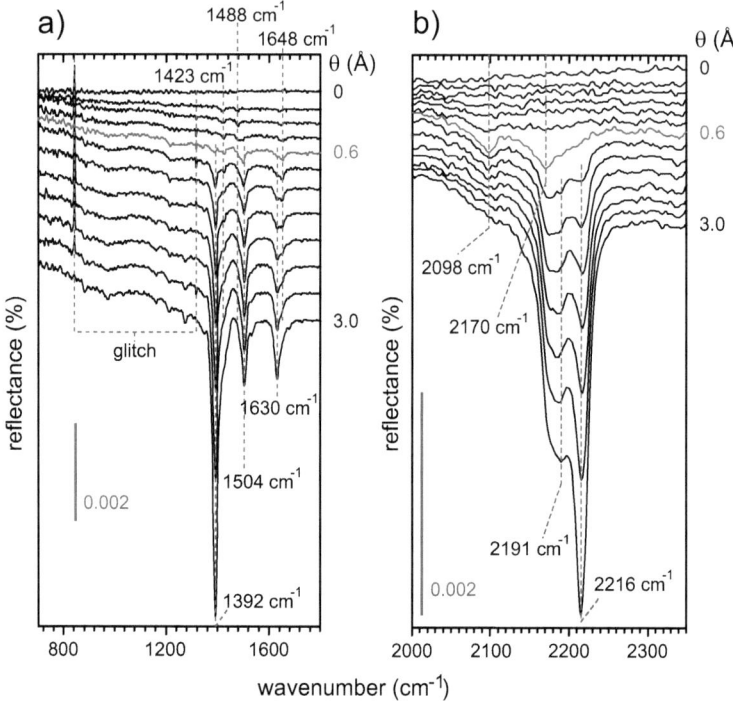

Figure 5.34.: RAIR spectra recorded during the deposition of F4-TCNQ on Ag(111) for coverages θ ranging from 0 to 5 Å. a) region from 700 to 1800 cm^{-1} and b) region of the CN stretch mode. For the most intense modes positions are given, for all other modes see Tab. A.6 in the appendix.

spectra in Fig. 5.34b, strong absorption is observed in the region of the ring deformation vibrations (1423, 1488, and 1648 cm^{-1}) and for the CN stretch modes (2098 and 2170 cm^{-1}) for sub-monolayer coverages of F4-TCNQ, which confirms the previous consideration. With increasing F4-TCNQ coverage, two new modes appear in the CN region, shown in Fig. 5.34b, which are attributed to the CN asymmetric (2191 cm^{-1}) and symmetric (2216 cm^{-1}) stretch mode of the neutral F4-TCNQ molecule. The two (sub-)monolayer CN stretch modes at 2098 and 2170 cm^{-1} are shifted to lower wavenumbers compared to the multilayer. This is an indication that the molecules become negatively charged and has also been reported for the parent molecule TCNQ [118, 184, 185]. This nicely corroborates the charge transfer from the metal to the molecule, which has been found in UPS measurements and the theoretical

5.4. The model acceptor F4-TCNQ adsorbed on Ag(111)

modeling. The modes at 1423, 1488, and 1648 cm^{-1} in the (sub-)monolayer spectrum in Fig. 5.34 are assigned to C=C stretching vibration of the ring (R in Fig. 5.33c and wing (W) of the F4-TCNQ molecule. This is in agreement with DFT calculations and the results of the parent molecule TCNQ[4] [185].

When the nominal coverage of F4-TCNQ is increased above the monolayer (0.6 Å), new peaks appear in the spectra, which can be attributed to the neutral F4-TCNQ molecule [186, 187]. Fig. 5.35 shows the full spectrum of a thick F4-TCNQ film on Ag(111) (nominally 20 Å). Note that the positions of the vibrational modes are not observed to shift when the coverage is increased from the 3 Å spectrum in Fig. 5.34 to the thick film spectrum in Fig. 5.35. However, the modes found in the multilayer are clearly different from the ones observed in the monolayer (some wavenumbers for the bulk vibrational modes are given in the spectra in Fig. 5.34, however a full list of all positions can be found in Tab. A.6 in the appendix). The reason being that the symmetry of the molecules in the bulk is *not* reduced and is thus D_{2h} as found for the molecule in the gas phase. This would in general reduce the number of allowed modes compared to the molecules in the monolayer (with point group C_{2v}) if the orientation of the molecules is unchanged. Since an increase in the number of modes is observed this leads to the conclusion that the molecules do change their orientation. Furthermore, the unit cell in bulk structure of F4-TCNQ is composed of four molecules [186], which can lead to a factor group splitting (Davydov-splitting) of some vibrational modes [188] additionally increasing the number of observed modes. As already noted, the asymmetric and symmetric stretching mode of the CN groups of neutral F4-TCNQ become visible in the spectra, which are in-plane of the molecule. This hints, that the molecules adopt an inclined orientation in the bulk in which the long and short molecular axes are tilted out of the substrate surface plane.

5.4.5. Conclusion

The strong electron acceptor F4-TCNQ has been adsorbed on Ag(111) and the evolution of its molecular features in the valence band and RAIR spectra have been monitored as a function of nominal layer thickness. Metal to molecule charge transfer was evidenced by the observation of a density of states in the energy gap region of the neutral molecule. Here, two states attributed to metal-molecule hybrid states derived from the former HOMO and LUMO of the neutral molecule are found. In addition, a linear increase in the work function by up to 0.65 eV (absolute value of $\Phi_{mod} = 5.20$ eV) is found in the SECO spectra. The experimentally found energy level diagram for F4-TCNQ is shown in Fig. 5.36. The results of the VB spectra are in good agreement with the adsorption of F4-TCNQ on Au(111) and

---

[4]The shift to higher wavenumbers compared to TCNQ is attributed to the influence of the fluorine atoms, compared to hydrogen in TCNQ.

Chapter 5. Results and Discussion

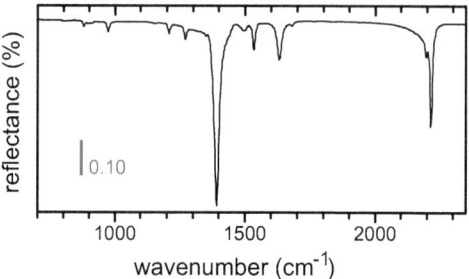

Figure 5.35.: RAIR spectrum of a nominal 20 Å thick film of F4-TCNQ on Ag(111) for the range from 700 to 2350 cm^{-1}

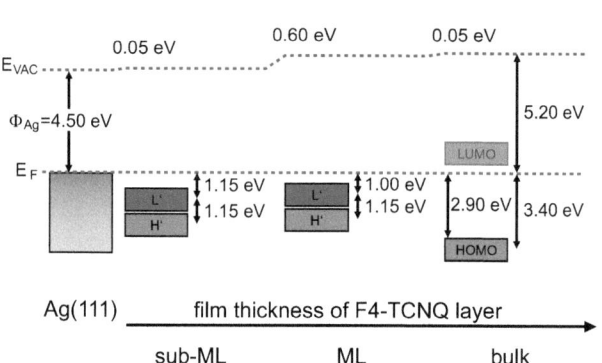

Figure 5.36.: Energy level diagram for F4-TCNQ adsorbed on Ag(111) as a function of nominal layer thickness from sub-monolayer coverage (sub-ML) to the bulk film. H' and L' denote metal-molecule hybrid states, E_F the Fermi level and E_{VAC} the vacuum level. Partially transparent boxes show molecular orbitals that have not been resolved experimentally.

Cu(111), which yielded a qualitatively similar picture. The experimental results are further backed by DFT modeling, which predicted an increase of the work function by +0.85 eV compared to the pristine metal. The underlying mechanism was electron transfer from the metal to the molecule and back donation from deeper lying molecular orbitals to the metal. The net charge transfer was calculated to 0.56 electrons per molecule, which is comparable to Cu(111) [16]. Due to the strong interaction with the substrate, the CN groups were found to bend down towards the surface resulting in a distorted adsorption geometry, again similar to results obtained theoretically and experimentally for F4-TCNQ adsorbed on Cu(111). The evaluation of the RAIR spectra showed that the CN stretching modes of F4-TCNQ molecules in the monolayer are red-shifted compared to the multilayers. This is a clear indication of negatively charged molecules in the (sub-)monolayer corroborating the findings of UPS and DFT. Moreover, the RAIRS data supports the assumption of the molecules adopting a face-on conformation in the first molecular layer. The weak attenuation of substrate features in the UP spectra indicates that island growth prevails in the multilayers. Furthermore, the orientation of the molecules in the multilayers is somewhat inclined having the long and short molecular axis tilted with respect to the surface plane.

5.5. Density dependent re-orientation and re-hybridization of chemisorbed conjugated molecules for controlling interface electronic structure

The adsorption of the molecular acceptor HATCN on Ag(111) was investigated in detail as function of layer density. It was found that the orientation of the first molecular layer changes from a face-on to an edge-on conformation depending on layer density, facilitated through specific interactions of the peripheral molecular cyano-groups with the metal. This is accompanied by a re-hybridization of molecular and metal electronic states, which significantly modifies the interface and surface electronic properties, as rationalized by theoretical modeling. This section was published as Ref. [189].

5.5.1. Introduction

Understanding the fundamental mechanisms that determine the properties of interface between metals and conjugated organic materials is a key prerequisite for advancing the fields of organic and molecular electronics, where the (opto-)electronic function of devices depends critically on, e.g., charge density distribution [190, 191], energy level hybridization and interface state formation [16, 190, 192], molecular conformation changes [16, 35, 192, 193], and energy level alignment [7, 35, 38, 194]. Additionally, it has been shown recently that the ionization energy of ordered molecular layers depends critically on the surface orientation of molecules, and that the energy levels at interfaces can be manipulated via molecular orientation [30]. However, it is commonly accepted that the orientation of a conjugated molecular monolayer with respect to a metal electrode surface is set for a particular material pair, only depending on the relative strength of metal-molecule and inter-molecular interactions. If substrate-molecule interactions prevail, as for clean metal surfaces, at least the first molecular layer is found to be face-on [50, 58, 65]. Only multilayers are known to eventually

Chapter 5. Results and Discussion

adopt a different orientation [55, 65, 70] because the strong interaction with the metal is already screened by the monolayer. Here, we demonstrate the molecular density-dependent re-orientation of an entire stable face-on monolayer to a stable edge-on conformation, which is achieved through balancing the surface energy in the two orientations by enabling strong hybridization of molecular and metal electronic states in both orientations through functional terminal groups. Particularly the chemisorbed molecular state in the edge-on monolayer exhibits extraordinary electronic properties with a potential for huge application relevance, which is not the case for the face-on orientation.

5.5.2. Evolution of the work function

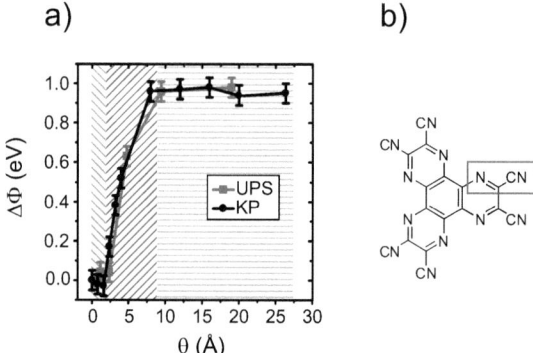

Figure 5.37.: a) work function change ($\Delta\Phi$) relative to pristine Ag(111) as a function of nominal HATCN film thickness θ, obtained by UPS (red squares) and KP (black dots). The full SECO spectra of the UPS experiments for the sequential deposition of HATCN on Ag(111) are shown in Fig. 5.43a. The color coding of the background shading represents the three different regimes referred to throughout the text: red = regime (i), black = regime (ii), and green = regime (iii). b) molecular structure of HATCN. The red rectangle shows a possible fragment in the TDS experiments with mass 52 g/mol.

The system under consideration is HATCN, a strong electron acceptor without an intrinsic dipole moment, chemisorbed on Ag(111). The Ag(111) surface work function (Φ) changes due to HATCN deposition as shown in Fig. 5.37a, determined independently by UPS and kelvin probe (KP) measurements (KP measurements have been done by Paul Frank, TU Graz, Austria). A pronounced deviation from the expected dependence of sample Φ on acceptor coverage (θ) is observed, as three regimes instead of two occur: (i) for low θ, Φ is essentially

5.5. HATCN on Ag(111)

constant (from 0 to ≈ 2.5 Å [5], highlighted in red in Fig. 5.37a), (ii) followed by a strong and almost linear Φ increase by ≈ 1 eV (from ≈ 2.5 Å to ≈ 8 Å, highlighted in black); (iii) finally a saturation of Φ is reached (beyond ≈ 8 Å, highlighted green). The value of Φ changes from 4.4 eV (pristine Ag(111)) to 5.4 eV (26 Å HATCN/Ag(111)). Normally, the deposition of acceptors on metal surfaces involving net metal-to-acceptor electron transfer leads to the observation of regime (ii) from the very beginning (compare Sec. 5.4), followed by regime (iii). Such a "delayed" increase of Φ, i.e., the additional appearance of regime (i), has not been reported before and cannot be rationalized on the basis of the Helmholtz-equation (or the Topping model including dipole-dipole depolarization effects [29]) when assuming any of the known growth models of conjugated molecules on metal surfaces (Sec. 2.2.1). As regime (ii) sets in after completion of a face-on nominal monolayer of HATCN [167], a considerable electron transfer from the metal also into HATCN multilayers would have to be invoked, which is inconsistent with the present understanding of organic/metal interface energetics [195]. The situation is further complicated by the fact that several possible orientations of molecules on top of the first face-on layer may occur. Hence, without exact knowledge of the orientation of HATCN molecules on Ag(111) and its evolution with θ, an explanation for Φ throughout regimes (i)-(iii) is impossible.

5.5.3. TDS and RAIRS experiments

The following TDS results have been obtained by Paul Frank (TU Graz, Austria). The quantitative analysis of TDS experiments for various initial HATCN coverages on Ag(111) is shown in Fig. 5.38, where the desorption of intact molecules (m = 384 g/mol) and HATCN fragments is plotted (C_2N_2, m = 52 g/mol; no other HATCN fragments were observed. One possible fragment of mass 52 g/mol is highlighted by the red rectangle in Fig. 5.37b. Alternatively, only the terminal CN groups might desorb as radicals and react to form C_2N_2 [196]. Consistent with the evolution of θ, Fig. 5.38 also displays three regimes: For coverages up to ≈ 2.5 Å no desorption of any HATCN fragments is observed throughout the entire accessible temperature range (up to 900 K). This is indicative of a HATCN layer strongly bonded to the Ag(111) surface. Between 2.5 and 8 Å, only desorbing C_2N_2 fragments are detected and virtually none of the molecules desorb intact. This shows that also in regime (ii) HATCN is strongly bonded to the metal, which would not be the case if they were part of a second layer. Instead, all molecules for θ up to 8 Å must be in direct contact with the Ag(111) substrate, and thus two different types of chemisorbed HATCN states (one for θ below 2.5 Å and one for thicknesses between 2.5 and 8 Å) exist. Beyond 8 Å an abrupt transition occurs: the TD signal intensity associated with fragments saturates and desorption of intact molecules

---
[5]The coverage values refer to the nominal mass thickness as read from the quartz crystal microbalance. 2.5 Å correspond to ca. one monolayer of face-on lying HATCN molecules.

Chapter 5. Results and Discussion

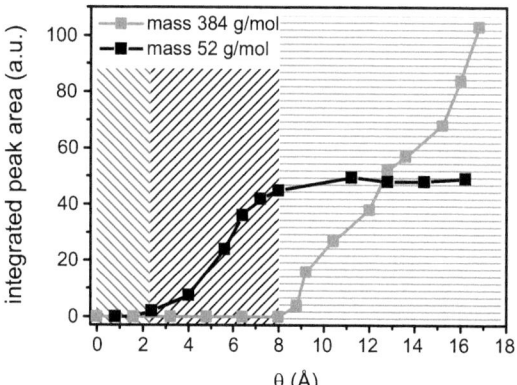

Figure 5.38.: TD spectra of mass 384 g/mol (HATCN) and mass 52 g/mol (HATCN fragment C_2N_2) for different θ on Ag(111). Data points were obtained by integrating the peak areas in the original TD spectra.

starts. The latter signal increases roughly linearly with θ and is thus associated with relatively weak bonded multilayers. The constant C_2N_2 signal implies that a HATCN layer equivalent to that of an initial θ of ≈ 8 Å exists below the multilayers.

The orientation and relative abundance of the two chemisorbed HATCN states in the first layer were further investigated by RAIRS. Here, the surface selection rules state that only those modes are observable, for which the dipole moment changes perpendicular to the surface as outlined in Sec. 3.2.2. HATCN in KBr serves as a reference for neutral molecules, and the region of the CN stretching vibration is shown in Fig. 5.39a, with a strong mode at 2241 cm^{-1} and a weaker one at 2251 cm^{-1}. This is fully consistent with the DFT calculated spectrum, which predicts an asymmetric and symmetric CN stretching mode, with the latter being at higher wavenumbers and having a lower absorption intensity as shown in Fig. 5.39a. The RAIR spectra recorded in-situ for sequential deposition of HATCN on Ag(111) are also shown in Fig. 5.39. Three characteristic peaks are observed, whose intensity changes as function of θ. From the very beginning a broad mode at 2185 cm^{-1} is seen (mode I), exhibiting a Fano-type lineshape (for a fit refer to Fig. A.8 of the appendix), which is reminiscent of dynamical charge transfer between substrate and molecule [197]. The strong shift of this CN stretching mode to lower energy compared to neutral HATCN indicates that the adsorbed molecule is negatively charged [198] similar to TCNQ and F4-TCNQ as discussed in Sec. 5.4. Mode I achieves its maximum intensity at $\theta \approx 2.7$ Å, where another mode at 2229 cm^{-1} (mode II) appears in the spectra. For higher coverages, the intensity of mode I decreases, while the intensity of

5.5. HATCN on Ag(111)

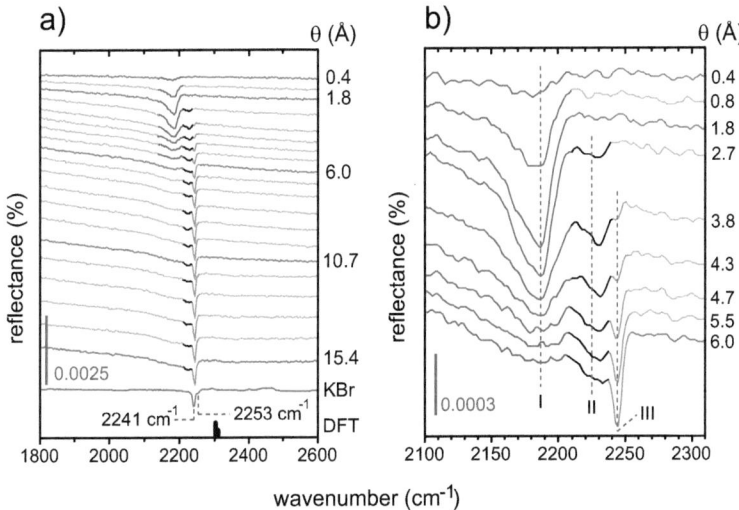

Figure 5.39.: a) RAIR spectra recorded during HATCN deposition on Ag(111) for the region of the CN stretch mode. b) Zoom into the region from 2100 to 2310 cm^{-1} for coverages from 0.4 to 6 Å. Labels I, II, and III represent the three different CN stretching contributions observed throughout the growth. Note that the KBr and the calculated spectrum have been scaled to fit the plot, besides the x-scale of the latter has been divided by 1.025 in order to correct for adsorption induced shifts as explained in Sec. 3.2.3. The color coding of the different spectral regions refers to the three regimes (i)-(iii) used in the text.

mode II saturates and can still be well resolved at $\theta = 15.4$ Å. Starting at $\theta = 3.8$ Å, a rather sharp mode at 2243 cm^{-1} (mode III) rises, which is the only one that continuously keeps increasing for higher θ. Mode I is attributed to face-on HATCN in direct contact with the Ag(111) surface, consistent with [167]. In this conformation the mode is IR active because of a significant dynamical charge transfer between molecule and surface and a non-planar conformation of the molecule (vide infra). The most striking observation is that this mode completely vanishes for $\theta \approx 8$ Å, which is another clear indication that the initial face-on HATCN chemisorbed state disappears. The intensity decrease of mode I begins ≈ 2.5 Å, similar to onset of the desorption of HATCN fragments in the TDS experiments. Mode II is associated with a HATCN state in which the molecules stand edge-on (possibly somewhat inclined) on the surface. In that case two of the CN groups are bonded to the surface, while the other four are only very weakly - if at all - are involved in the molecular interaction with

Chapter 5. Results and Discussion

the surface. The smaller energy shift of mode II relative to the mode of neutral HATCN in KBr compared to mode I results from the bond weakening due to charge transfer to the CN groups (chemical shift, as described in Sec. 3.2.3), which is partially compensated by C and N atoms now oscillating "between" the Ag surface atoms and the rest of the molecule (mechanical renormalization). As mode III is very sharp and its position is close to modes of HATCN in KBr it is attributed to originate from multilayer HATCN, possibly superimposed by the vibrations of the non-bonded CN groups in the now edge-on oriented monolayer.

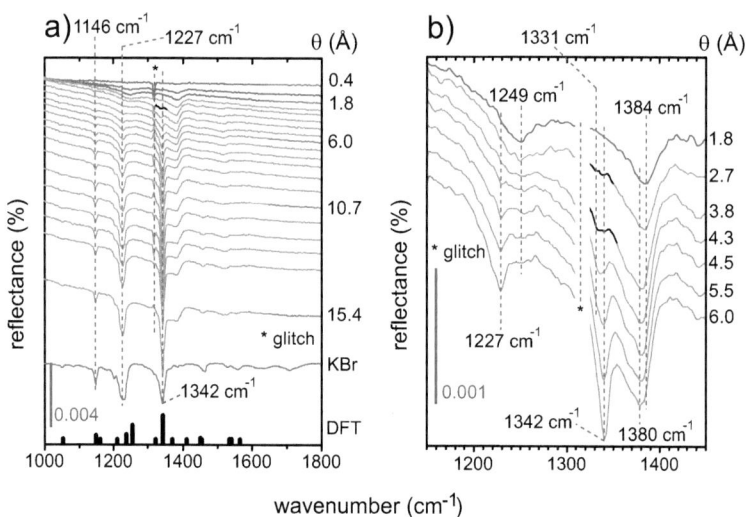

Figure 5.40.: a) RAIR spectra recorded during HATCN deposition on Ag(111) for the region between 1000 and 1800 cm^{-1}. Besides the spectrum of HATCN pressed in a KBr disc and the DFT calculated modes for the neutral molecule are presented in a). Note that the KBr and the calculated spectrum have been scaled to fit the plot, besides the x-scale of the latter has been divided by 1.025 in order to correct for adsorption induced shifts as explained in Sec. 3.2.3. b) Zoom into the region from 1150 to 1450 cm^{-1} for coverages from 0.4 to 6 Å. The color coding of the different spectral regions refers to the three regimes (i)-(iii) used in the text.

RAIR spectra of the region from 1000 to 1800 cm^{-1} are shown in Fig. 5.40 for film thicknesses ranging from 0 to 15.4 Å. The colors of the different modes in the spectra have been chosen to fit with the assigned regimes (i.e. regime (i) = red, (ii) = black and (iii) = green) as described before. At the coverage corresponding to a nominal thickness of 1.8 Å coverage, two broad modes can clearly be observed at ≈ 1249 and ≈ 1384 cm^{-1} (as shown in Fig. 5.40a and b).

5.5. HATCN on Ag(111)

They possibly belong to totally symmetric in-plane modes, that become IR-active due to the interaction (conformational change as shown in Fig. 5.42 and possibly dynamical charge transfer) of face-on HATCN with the Ag(111) substrate [113]. At the onset of regime (ii), both modes start to vanish and another mode at 1331 cm^{-1} appears (marked black in the spectra corresponding to 2.7 and 3.8 Å, as shown in Fig. 5.40a and b). This mode is attributed to the molecules starting to assume an edge-on configuration in regime (ii). With increasing coverage it becomes hidden under the strong modes of the neutral HATCN molecules in the bulk. The decrease of the two modes of regime (i) indicates that the face-on molecules are starting to re-orient and become incorporated into an edge-on molecular monolayer. With increasing HATCN deposition all modes of the neutral molecule [198] appear in the spectrum (green spectra in Fig. 5.40a and b compared to the blue HATCN KBr spectrum). The strongest ones are located at 1146, 1227 and 1342 cm^{-1} and are assigned to in-plane breathing modes of the HAT-backbone by DFT gas phase calculations. They can only be observed if the molecule is in an upright conformation and are thus attributed to regime (iii).

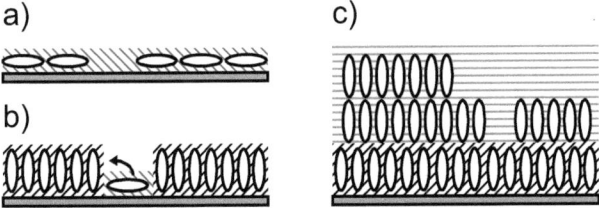

Figure 5.41.: Proposed density dependent re-orientation of HATCN on Ag(111) and Cu(111). The color coding of the background shading represents the three different regimes referred to throughout the text: a) red = regime (i), b) black = regime (ii), and c) green = regime (iii).

The (re-)orientation of HATCN in the first few layers is schematically summarized in Fig. 5.41, where three regimes can be distinguished, as was already the case in the initial observation of the work function change in Fig. 5.37a. Therefore, a direct link between orientation of the molecules and Φ can be established: In the initial stage of film growth (from 0 Å to ≈ 2.5 Å), the molecules are lying face-on and Φ remains constant. With further deposition, the chemisorbed face-on HATCN molecules re-orient and are incorporated into a monolayer, where the molecules are aligned edge-on in a different chemisorption state (This growth classification is also supported by Auger electron spectroscopy experiments, in which the intensity decrease of an Ag auger transition was monitored during growth as shown in Sec. A.9 of the appendix). This particular re-arrangement leads to the large Φ increase until saturation is reached at the end of regime (ii). It should be emphasized that the re-orientation

Chapter 5. Results and Discussion

cannot be induced by thermal annealing, and that the face-on chemisorbed layer is stable for several days as long as no HATCN is further deposited. Experiments for HATCN deposited on Cu(111) substrates yielded fully analogous results in terms of re-orientation and work function change regimes, which will be shown in Sec. 5.6.

5.5.4. Theoretical modeling of HATCN on Ag(111)

The following theoretical results have been obtained by Oliver T. Hofmann (TU Graz, Austria). To understand the mechanisms for the behavior of Φ, the face-on and edge-on HATCN layers on Ag(111) were modeled using the VASP code and DFT methodology as outlined in Sec. 4.5. For the face-on conformation (assuming the structure from [167]) of regime (i) it is found that the CT significantly outweighs Pauli repulsion, giving a Φ change of $\Delta_{CT} = +0.7$ eV. However, since the CN groups at the periphery of HATCN bend down towards the surface (Fig. 5.42b and Ref. [167]), a dipole in the opposite direction is created, yielding a molecular conformation related Φ change $\Delta_{Mol} = -0.5$ eV. As a net result a small Φ increase of +0.2 eV is obtained by DFT. For the edge-on conformation of HATCN in regime (ii), no experimental evidence for an ordered molecular layer could be provided yet, which points to a disordered layer or a very complex surface unit cell (as already found for bulk HATCN *vide supra*). Thus, a single upright standing molecule in a $2 \times 3\sqrt{3}$ unit cell (for details see Sec. A.8 of the appendix) was chosen for the calculations. In this conformation, geometric distortions of the molecule are small and Δ_{Mol} falls below +0.1 eV. The mechanism of the CT between metal and molecule is altered fundamentally by the re-orientation. Fig. 5.42a and b show the charge rearrangements upon adsorption of HATCN in the two orientations. For the face-on conformation in regime (i) the whole molecule is involved in the interaction with the substrate, including π-electrons of the CN groups as well as the entire π-system (As proved experimentally by the observation of totally symmetric modes in the region of the in-plane ring deformation and the CN vibrations). For the edge-on conformation, the CT becomes more localized on the "docking groups" (here: CN), similar to typical self-assembled monolayers [199]. The charge redistribution on the molecular backbone only plays a minor role. Yet the charge rearrangements occur over a larger distance than in the face-on conformation. In conjunction with the increased packing density and the vanishing Δ_{Mol}, this results in an increased Φ of +2.4 eV. The results for both regimes (i) and (ii) are in reasonable agreement with the experimental observations, as they show that the Φ change of the face-on monolayers is negligible compared to that induced by the edge-on layer. A better (i.e., quantitative) correlation between theory and experiment cannot be expected, particularly considering the unknown structure of the edge-on layer.

5.5. HATCN on Ag(111)

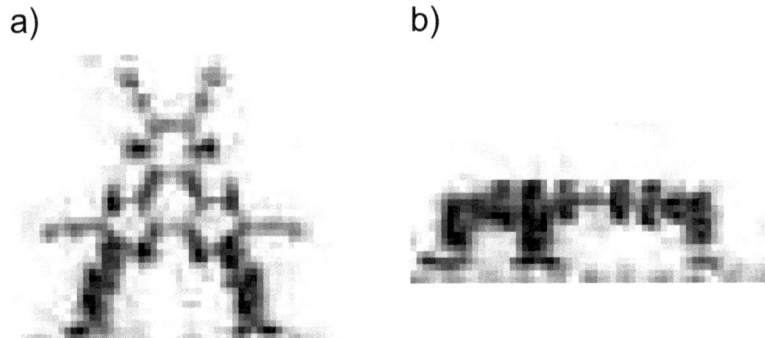

Figure 5.42.: Charge rearrangements upon adsorption in the edge-on a) and face-on b) HATCN conformation as inferred from DFT. Red indicates increased electron density, blue electron density depletion.

5.5.5. UPS valence band spectra of HATCN on Ag(111)

The reason why the Φ increase sets in only in regime (ii) can be understood based on charge rearrangement and molecular distortion related arguments as described before. From an energy level alignment point of view it appears, however, much less obvious at first sight. The experimental UPS spectra in Fig. 5.43 have been recorded during the sequential deposition of HATCN on Ag(111). As shown in detail in the lower part of Fig. 5.43c, new intensities appear in the energy gap of the neutral molecule close to the Fermi level (denoted with L' and H'), which are ascribed to hybrid metal-molecule states derived from the former LUMO (L') and HOMO (H') level of neutral HATCN. No obvious differences in the shape of these two features can be found for HATCN adsorbed face-on (0 to ≈ 2.5 Å) or edge-on (≈ 2.5 to ≈ 8 Å), which is further supported by DFT calculations shown in the upper part of Fig. 5.43c. Here, the LUMO level of the bonded HATCN molecules becomes partially filled also in both conformations (filling is indicated by the red area in the upper part of Fig. 5.43c), and the metal Fermi level is consequently pinned at the (former) LUMO. As a result, the shift of the potential landscapes in the metal and the molecular layer and the Φ modifications in both cases have to correspond to the difference between the EA of the adsorbate (i.e., the corresponding LUMO energies) and Φ of the bare substrate. Thus, one might come to believe that $\Delta\Phi$ should be (almost) independent of the adsorption geometry. This line of arguments, however, misses an essential property of molecular layers described by Duhm and coworkers [30], namely that also in molecular systems the EA is a quantity that depends on the direction in which an electron leaves an oriented layer. Indeed, it is found that the EA

Chapter 5. Results and Discussion

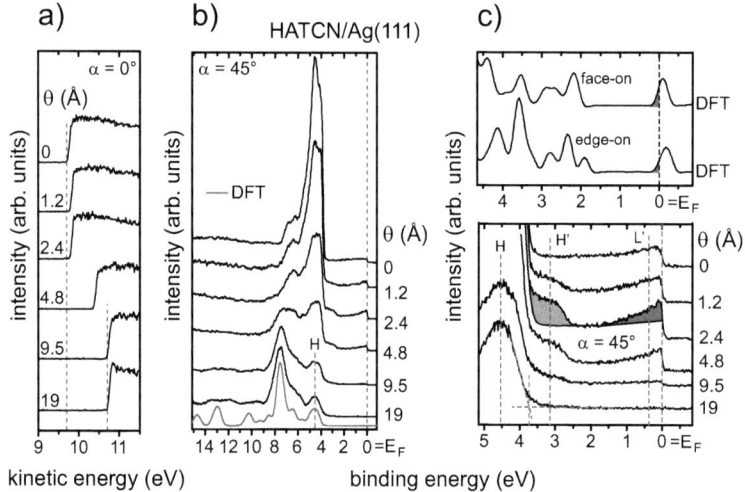

Figure 5.43.: The valence band electronic structure of the sequential deposition of HATCN on Ag(111) as obtained by UPS is shown in a), b) and the lower part of c). Furthermore, the DFT computed densities of states projected onto the HATCN layer are plotted in the upper part of c). The red area represents the part of the metal-molecule hybrid state (derived from the former LUMO of neutral HATCN) that is partially filled upon adsorption for both conformations (L'). H' denotes the metal-molecule hybrid state derived from the former HOMO, H the HOMO of the (neutral) bulk film, and E_F the Fermi level. For comparison, the DFT calculated gas phase DOS of HATCN is plotted in b) (applied Gaussian broadening with FWHM = 0.5 eV and shifted in x-direction).

perpendicular to the sample surface for face-on and edge-on planar HATCN molecules differs by 1.15 eV. The difference in $\Delta\Phi$ is further increased by the distortion-induced dipole of the adsorbed face-on molecules (denoted Δ_{Mol} vide supra). The HOMO of neutral HATCN in the bulk (denoted H in Fig. 5.43b and c) is found at 4.55 eV BE in the 19 Å spectrum. The onset is located at 3.70 eV BE yielding an ionization energy of $IE_{Ag,HATCN} = 9.10$ eV, which compares well with the theoretically predicted gas phase value of $IE_{HATCN}^{vert} = 10.40$ eV (bearing in mind that the IE of molecules in a bulk film is generally reduced compared to the gas phase due to polarization effects [200]). Additionally, the DFT calculated DOS of HATCN in the gas phase (Gaussian broadening with FWHM = 0.5 eV) is shown in Fig. 5.43b (it has been stretched and shifted to fit the experimental spectrum), which nicely reproduces the experimental intensities.

5.5.6. Hole injection barrier decrease for α–NPD by HATCN interlayers

One example for hole injection barrier lowering using an edge-on sub-monolayer of HATCN on Ag(111) is shown in Fig. 5.44. Here, the modified work function of the substrate is 5.3 eV (5 Å HATCN). The obtained Δ_h into subsequently deposited 30 Å of α–NPD is $\Delta'_h = 0.4$ eV, which is a decrease by 1 eV compared to Δ_h for α–NPD on pristine Ag(111) of $\Delta_h = 1.4$ eV [201]. Furthermore, a vacuum level shift occurs between HATCN and α–NPD leading to a work function of 4.95 eV after deposition of the hole transport material. This shift is rationalized by a pinning of the molecular HOMO of α–NPD at the Fermi level ($IE_{\alpha-NPD} = 5.2$–5.5 eV see Sec. 4.1.2).

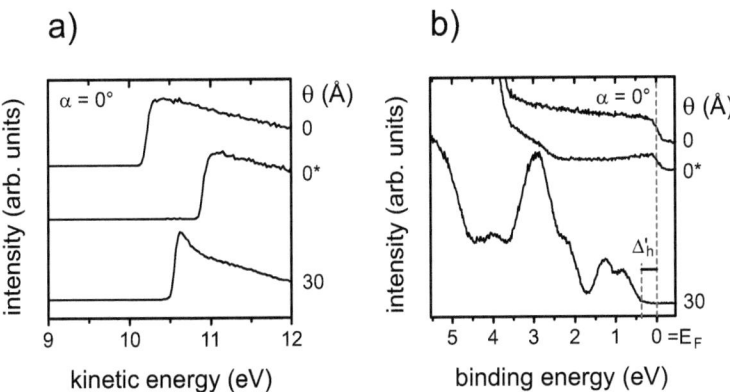

Figure 5.44.: SECO (a) and valence band (b) spectra of the deposition of 30 Å α–NPD on HATCN modified Ag(111). The nominal thickness of the HATCN layer (labeled "0*") was 5 Å. Δ'_h denotes the hole injection barrier and E_F the Fermi level.

5.5.7. Conclusions

The sequential deposition of HATCN on Ag(111) has been studied using PES, TDS, and RAIRS. The strong work function change of up to 1 eV was observed to occur in three regimes depending on the molecular density. Via TDS and RAIRS experiments, the three regimes could be linked to face-on molecules in regime (i) (minor change in Φ), a gradual re-orientation of the face-on molecules to an edge-on orientation in regime (ii) (strong work function increase), and finally a saturation of the work function at an absolute value of 5.6 eV. The microscopic origin of the density dependent molecular re-orientation must be sought in a subtle interplay of various effects. The σ–bonding strength between the CN groups and

Chapter 5. Results and Discussion

the metal through the N lone pair can be expected to be larger for edge-on molecules; at low θ, it is, however, still favorable for HATCN to adsorb in a face-on geometry bringing all six CN groups into contact with the metal, and maximizing the interactions with the substrate while at the same time minimizing the metal surface energy. Once the number of molecules is larger than needed to close the face-on layer, the total number of CN groups in contact with the metal and their respective bonding strength is increased when at least part of the molecules change to edge-on. Thus, the presence of the functional groups at the periphery of HATCN facilitates the re-orientation of the layer, while in their absence the second layer would just grow in a different orientation on the unchanged face-on monolayer. In our specific example, the structural change of the molecular monolayer was accompanied by a significant change in surface functionality, i.e., Φ for the edge-on layer was 1 eV higher than for the face-on layer or pristine Ag. In comparison, a monolayer of the widely used strong acceptor tetrafluoro-tetracyanoquinodimethane could increase Φ of Ag surfaces by only 0.65 eV (see Sec. 5.4). This renders the high Φ surface an interesting candidate as hole injecting electrode in electronic devices, as the absolute work function value of 5.6 eV parallels that of atomically clean Au, however, the electron "push-back" effect, which is detrimental for achieving low hole injection barriers on clean metal surfaces [191, 38] is expected to be minimized by the surface termination with the molecule. This has been shown for the hole transport material $\alpha-$NPD, where a lowering of Δ_h of 1 eV was obtained. Also, it might be interesting to evaluate whether interface dipoles reported for other organic/metal interfaces involve in part a contribution from a multilayer-induced re-orientation of the buried monolayer in direct contact with the metal.

5.6. HATCN adsorbed on Cu(111) and Au(111)

In addition to the adsorption of HATCN on Ag(111) as described in the previous section, HATCN deposited on Cu(111) and Au(111) has been thoroughly investigated by UPS and RAIRS. On copper, a similar density depended re-orientation of the first molecular layer of HATCN molecules is found as on Ag(111). Differences are observed in the structure of the first molecular layer: on Ag(111) HATCN molecules pack in a hexagonal superstructure [167], whereas a dendridic/fractal-like structure prevails on Cu(111). The interaction between the Au(111) substrate and HATCN is rather weak, which promotes the formation of 3D islands. Nevertheless, the first HATCN layers are originally oriented parallel to the surface, but re-orient when the density of HATCN molecules is increased, too. Part of this section will be published in a forthcoming publication.

5.6. HATCN adsorbed on Cu(111) and Au(111)

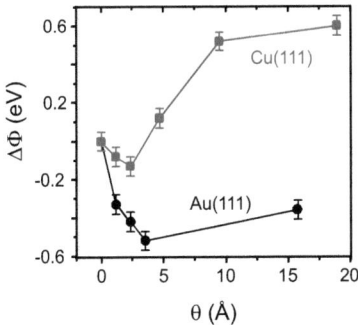

Figure 5.45.: Evolution of the work function change ΔΦ as a function of coverage θ for the sequential deposition of HATCN on Cu(111) (black) and Au(111) (red).

5.6.1. Valence electronic structure

The evolution of the work function change ΔΦ as a function of HATCN coverage θ on Cu(111) and Au(111) is plotted in Fig. 5.45. On copper, a qualitatively similar evolution of the work function as reported in the previous section for HATCN on Ag(111) is found. On Cu(111), the work function decreases by 0.15 eV from 4.95 eV (pristine Cu(111)) to 4.80 eV (2.4 Å HATCN coverage) in regime (i). In regime (ii) Φ is observed to increase almost linearly by 0.65 eV until an absolute Φ value of 5.45 eV at nominally 9.5 Å is reached. Further increase in coverage leads to saturation of the work function (regime (iii)) and only a minor increase to 5.55 eV is observed. Similar to the argumentation on Ag, the initial slight decrease in the work function is attributed to a CT between metal and molecule, which partially compensates for the "pushback" effect. In the UP VB spectra (shown in Fig. 5.46b and c) again similar observations compared to the adsorption of HATCN on Ag(111) are made: Due to the CT, metal-molecule hybrid states are created in the energy gap region of the neutral molecule. One state, which is located closest to the Fermi-edge, is derived from the former LUMO (L' in Fig. 5.46c) and the other at slightly higher BE from the former HOMO (H') of neutral HATCN. However, in contrast to Ag(111), it seems that the L' state is located closer to the Fermi level for thin films compared to coverages of 7.1 and 9.5 Å, where it is found at higher BE (≈ 0.5 eV BE, denoted L" in Fig. 5.46b). The reason behind this might be a different hybridization between the metal and face-on/edge-on HATCN layer. This issue will be addressed in a further theoretical study. Similar to the case of 3,4,9,10-perylenetetracarboxylicdianhydride (PTCDA) adsorbed on coinage metal substrates ([36]), where the LUMO derived hybrid state is located at higher BE for a larger CT, similar arguments might hold for the two states (L' and L") observed for HATCN on copper. The HOMO of neutral HATCN in the bulk (H) is

Chapter 5. Results and Discussion

located at 4.60 eV with its onset at 3.80 eV in the 19 Å spectrum, which yields an ionization energy of $IE_{Cu,HATCN} = 9.35$ eV. This is in good agreement with the value obtained on silver and the calculated IE.

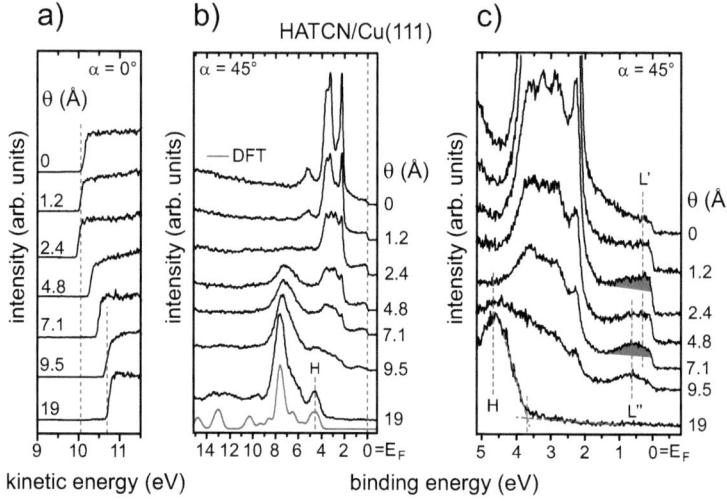

Figure 5.46.: The valence band electronic structure of the sequential deposition of HATCN on Cu(111)as obtained by UPS is shown. The red area in c) represents the part of the metal-molecule hybrid state (derived from the former LUMO of neutral HATCN) that is partially filled upon adsorption for thin films (L') and thicker films (L"). H' denotes the metal-molecule hybrid state derived from the former HOMO, H the HOMO of the (neutral) bulk film, and E_F the Fermi level. For comparison, the DFT calculated gas phase DOS of HATCN is plotted in b) (applied Gaussian broadening with FWHM = 0.5 eV and shifted in x-direction).

On Au(111) the VB spectra of the sequential adsorption of HATCN are plotted in Fig. 5.47b and c. From the absence of metal-molecule hybrid states in the energy gap region of HATCN, it can be concluded that the interaction between gold and HATCN is weaker compared to Ag and Cu. This is further evidenced by the slower attenuation of photoemission features compared to both other substrates, since even at a nominal coverage of 16 Å the Au 5d and 6s features (Fermi-edge) are still visible (Fig. 5.47b and c). By comparison of the 16 Å spectrum and the theoretically obtained HATCN gas phase DOS (red spectrum in Fig. 5.47b), the position of the HOMO feature (H) can be determined centered at ≈ 4.60 eV BE. Using the same energetic difference between peak maximum and onset as obtained on Cu(111), the

5.6. HATCN Adsorbed on Cu(111) and Au(111)

onset is located at ≈ 3.80 eV BE. Weak interaction between HATCN and Au(111) is also reflected in the evolution of the work function change $\Delta\Phi$, which is shown in Fig. 5.45. Here, a maximum decrease by −0.5 eV from 5.50 eV (pristine Au(111)) to 5.00 eV at a coverage corresponding to about one face-on monolayer (≈ 3 Å, taking the change in Φ for HATCN on Ag(111) and Cu(111) as a reference for one completed face-on monolayer) is observed. With further increasing the coverage, a slight increase in the work function is obtained (+0.15 eV). The initial work function drop can be rationalized again by two counteracting mechanisms: (i) weak chemical interaction between substrate and HATCN which partially compensates the (ii) "push-back" effect (Φ decreases beyond 1 eV are usually reached on clean Au(111)). The work function increase for higher coverages can be explained by a change in the structure of the adsorbed HATCN film, which will be discussed together with the RAIRS results (vide infra). Taking the absolute value for the work function of the 16 Å HATCN film (5.15 eV) and the approximated onset into account an ionization energy of $IE_{Au,HATCN} = 8.95$ eV is obtained.

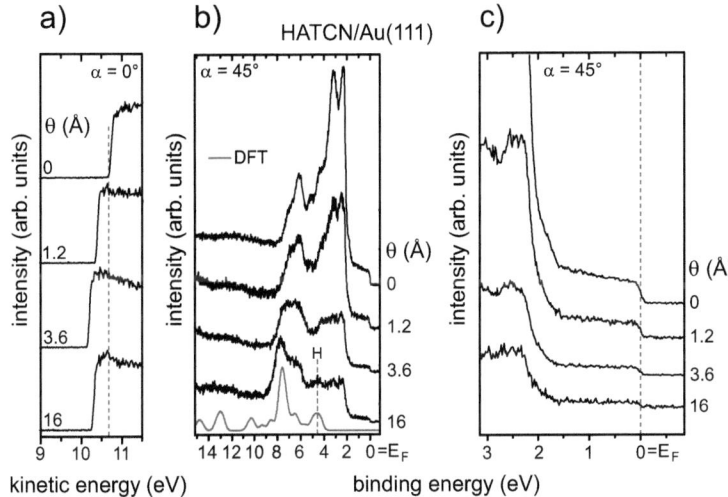

Figure 5.47.: The valence band electronic structure of the sequential deposition of HATCN on Au(111) as obtained by UPS is shown. a) SECO spectra, b) full spectrum of the valence region, and c) zoom into the near Fermi level region of the valence region. In b) also a DFT computed gas phase DOS of HATCN is plotted (applied Gaussian broadening with FWHM = 0.5 eV and shifted in x-direction). H denotes the HOMO of the (neutral) bulk film and E_F the Fermi level.

5.6.2. RAIRS results

The RAIR spectra for the deposition of HATCN on Cu(111) and Au(111) recorded in-situ during growth from 0 to 20 Å nominal film thickness are plotted in Figs. 5.48 (Cu) and 5.49 (Au). On Copper, similar observations for the evolution of the vibrational modes with coverage are made in the region of the in-plane ring deformation region (shown in Fig. 5.47a) compared to Ag(111). However, only one mode at 1375 cm^{-1} is observed for coverages up to 5.1 Å, which might point to a slightly different conformation and/or CT. This mode must be assigned to a totally symmetric vibration, since the molecules are observed to adopt a face-on conformation on Cu(111) by STM (similar to Ag(111), however with a different arrangement as shown in Fig. A.10 of the appendix). Between 1.6 and 5.1 Å a slight blue-shift to 1382 cm^{-1} can be observed. Increasing the coverage of HATCN on the surface beyond 5.1 Å leads to the appearance of new modes with the most intense ones at 1147, 1225, and 1342 cm^{-1}. These positions are almost exactly the ones observed on Ag(111) and furthermore their intensity ratio is similar. Consequently, they are attributed to HATCN molecules in the bulk, which adopt a similar (upright) structure as on Ag(111).

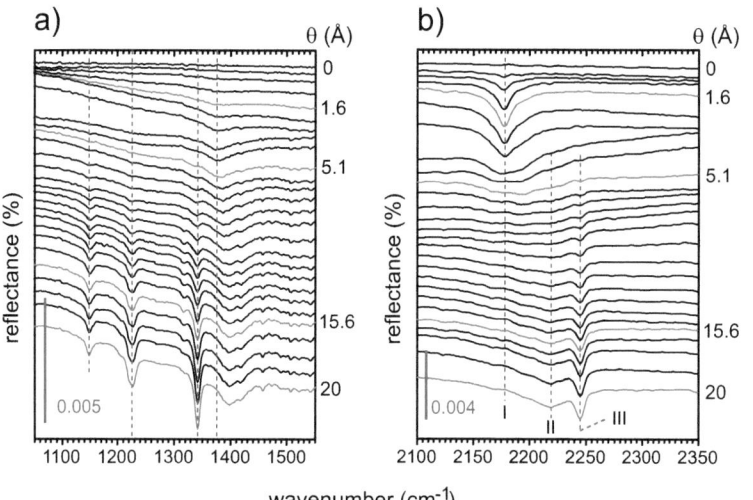

Figure 5.48.: RAIR spectra for the deposition of HATCN on Cu(111) recorded in-situ. a) region of the in-plane ring deformation vibrations and b) region of the CN stretching mode. For the description of the modes labeled I, II, and III see text.

5.6. HATCN adsorbed on Cu(111) and Au(111)

In the region of the CN stretching vibrations (Fig. 5.48b), three characteristic peaks are observed, whose intensity changes as function of θ. From the very beginning a broad, but symmetric mode at 2177 cm^{-1} is seen (mode I). Its shape is characteristic for the HATCN molecules forming a 2D lattice gas [106], which is consistent with the structure observed in the STM images (Fig. A.10). Increasing the coverage beyond 1.6 Å leads to a strong broadening and blue-shift of mode I. Additionally, it is observed to decrease in intensity and finally vanishes for coverages of $\approx$ 8 Å, while two other modes appear in the spectrum (mode II and III). Mode II is observed at 2219 cm^{-1} and initially increases in intensity until it saturates at coverages between 15 and 20 Å. Mode III is located at 2244 cm^{-1} and rather sharp compared to the other two modes. It is the only mode that continuously growths with increasing deposition and is thus attributed to molecules in multilayers. The observed evolution of the vibrational modes in the CN region for HATCN on Cu(111) is thus in full agreement with that on Ag(111). Consequently, a very similar re-orientation of the first face-on monolayer to an upright edge-on layer with increasing density of HATCN molecules as shown in Fig. 5.41 takes place on Cu(111).

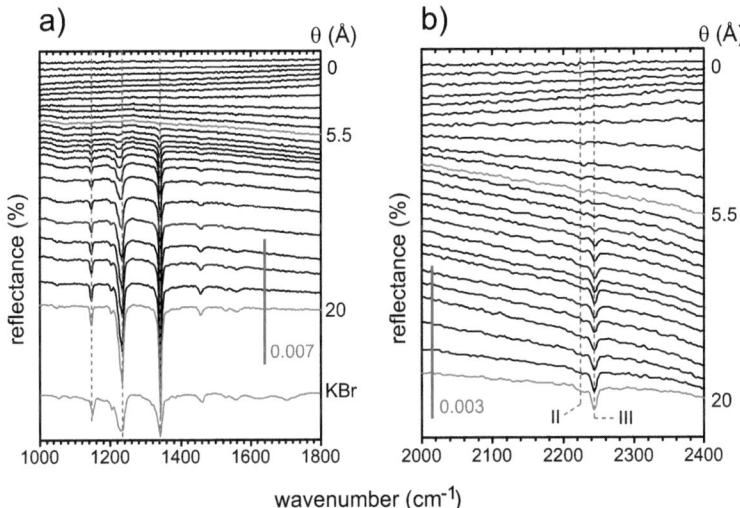

Figure 5.49.: RAIR spectra for the deposition of HATCN on Au(111) recorded in-situ. a) region of the in-plane ring deformation vibrations and b) region of the CN stretching mode. A spectrum of HATCN pressed in a KBr disc is also included for the region in a). It has been scaled in intensity to fit the plot. For the description of the modes labeled II and III in b) see text.

Chapter 5. Results and Discussion

The RAIR spectra recorded for the deposition of HATCN on Au(111) are shown in Fig. 5.49. In both regions (in-plane ring deformation Fig. 5.49a and CN stretching vibrations Fig. 5.49b) vibrational modes are not observed until a nominal coverage of 5.5 Å is reached (deposition with the sample held at -155 °C show molecular signatures right from the beginning). This indicates that the HATCN molecules on the surface must be oriented face-on as even only a slight tilt would make the in-plane modes of HATCN allowed by the surface selection rule. Additionally, the interaction with the substrate must be rather weak and as a consequence no totally symmetric mode becomes observable as the corresponding dynamical charge transfer between metal and molecule is too weak. Increasing the coverage beyond 5.5 Å leads to the observation of the modes of the neutral HATCN molecule as the comparison with the thick film spectra on Ag(111) and Cu(111) and the KBr spectrum shows. In the CN region (shown in Fig. 5.49b) two modes are observed (labeled mode II and III). Mode III is located at 2244 cm^{-1} and starts to increase for coverages beyond 5.5 Å, which is in agreement with the other modes attributed to HATCN adopting the bulk structure. For coverages between 5.5 and 20 Å another mode (denoted II in Fig. 5.49b) appears in the spectra at 2226 cm^{-1}, which is attributed to edge-on HATCN molecules in direct contact with the substrate in agreement with the results obtained on Ag(111) and Cu(111). Its lower wavenumber compared to the mode III of the bulk indicates chemical interaction between the substrate and the molecule, possibly induced by weak CT, since the work function in Fig. 5.45 also increases slightly for coverages above 5 Å. The molecular growth of HATCN on Au(111) has been described in a recent study by Frank and coworkers [202] and is schematically depicted in Fig. 5.50. Using this study as reference, the RAIRS results will be discussed in the following. For sub-monolayer coverages, the HATCN molecules adsorb in an face-on orientation without considerable lateral interactions (Fig. 5.50a). Starting from the clean surface, the first molecular layer is filled up to a certain saturation coverage with continuous deposition of HATCN (not necessarily corresponding to a full monolayer). Then, a second face-on layer starts to built up on top of the first one, as shown in Fig. 5.50b. This is consistent with the RAIRS results, where no change in reflectance is observed up to a coverage of 5.5 Å, which is twice the amount that has been attributed to a face-on monolayer of HATCN on Ag(111) ($\approx$ 2.5 Å). Further increase of HATCN deposition on the surface leads to the formation of molecules in the bulk structure with an upright orientation (Fig. 5.50c) and consequently the modes of the neutral molecules in the bulk structure are observed in the RAIR spectra. In this bulk structure also molecules from the second face-on layer are incorporated as obtained in the experiments by Frank and coworkers. With more and more molecules arriving at the surface, higher 3D islands are build and also the face-on molecules in the first layer are re-oriented to an edge-on conformation and incorporated into the islands as shown in Fig. 5.50d and e. This is also corroborated by the advent of mode II in the RAIRS experiments, which evidences the formation of edge-on

5.6. HATCN adsorbed on Cu(111) and Au(111)

molecules in direct contact with the substrate. The 3D island growth of HATCN on Au(111) is supported by the slow attenuation of the substrate's photoemission features in the VB in Fig. 5.47.

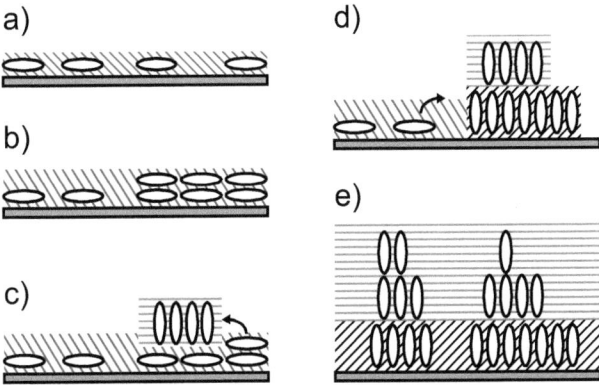

Figure 5.50.: Density dependent re-orientation of HATCN on Au(111) as proposed by Frank and coworkers [202]. The color coding of the background shading represents face-on molecules in the first and second layer (red), edge-on molecules in contact with the metal substrate (black), and molecules in the bulk (green).

5.6.3. Conclusion

HATCN has been adsorbed on Cu(111) and Au(111) metal single crystals and the evolution of the electronic and structural properties has been followed using UPS and RAIRS. In both cases, a re-orientation of initially face-on molecules in the first molecular layer(s) to edge-on molecules is found, similar to the case of Ag(111) presented in Sec. 5.5. On copper, the structure of the first molecular layer is found to be randomly ordered compared to Ag(111), where a well ordered superstructure has been observed. The re-orientation takes place after (almost) completion of the first face-on monolayer, as evidenced by the coverage dependent RAIR spectra. Further support for the re-orientation comes from work function measurements. Here, the evolution of Φ as a function of coverage shows three regimes, similar to the results on Ag(111). However, the maximum work function increase on copper is 0.6 eV compared to 1.0 eV on silver. The valence band spectra show additional density of states in the energy gap of the neutral molecule, which is assigned to metal-molecule hybrid states derived from the former LUMO and HOMO of the neutral molecule. In contrast to the results on silver, the hybrid state derived from the former LUMO is located at higher BE after

Chapter 5. Results and Discussion

the re-orientation. Besides the filling of the LUMO observed in the VB, further support for the CT from the metal to the molecule comes from the RAIRS data. Here, a red-shift of the CN stretching vibration of molecules in the monolayer compared to the (neutral) multilayer is observed. The proposed growth of HATCN on copper is similar to that on silver as shown in Fig. 5.41. On gold weak interaction between HATCN and substrate is supported by an initial work function decrease upon adsorption. This decrease is mainly due to the electron "push-back" effect, however, not exclusively as the work function drop is only 0.5 eV. In addition, the work function increases again after a certain amount of molecules has been deposited on the surface. This can be attributed to the re-orientation of the molecules and weak chemical interaction between the edge-on molecules and the substrate. However, no additional states appear in the energy gap region of the neutral molecules in the VB for all coverages. The RAIR spectra of the CN stretching vibrations show that approximately the first two molecular layers are oriented initially face-on, since no intensity/reflectance change is observed. This further underlines the weak interaction between molecules and surface, as no dynamical charge transfer dipoles are created, which would be observable even though the molecules are oriented face-on. Increasing the HATCN coverage further leads to the appearance of molecular derived features in the RAIR spectra evidencing edge-on molecules. The re-orientation can thus not be evidenced from the RAIRS data alone, but has been observed in a recent TDS study [202]. There it was found that the re-orientation occurs after the second layer is build up, however slightly more complex compared to silver and copper as shown in Fig. 5.50.

5.7. Tuning the hole injection barrier from indium tin oxide into subsequently deposited α−NPD by electron acceptor pre-coverage

5.7.1. Introduction

Charge injection from electrodes into subsequently deposited organic layers is an extremely important process in organic devices as outlined in Sec. 1. ITO is still one of the most frequently used anodes to inject holes into organic layers in OLED devices. However, often a large hole injection barrier (Δ_h) is found at the ITO/organic interface because the work function of ITO is in the range of 4.1−4.7 eV (depending on its pretreatment), which is significantly lower than the typical ionization energies of organic materials [203]. To increase the work function of ITO and to decrease Δ_h into organic hole transport materials, surface pre-coverage with molecular electron acceptors was explored. It will be shown in the following that the work function of ITO substrates can be strongly increased depending on the acceptor pre-coverage (similar to coinage metal surfaces). Furthermore, linear tuning of the hole injection barrier from ITO into α−NPD up to a minimal value $\Delta_{h,F4-TCNQ}^{min} \approx 0.9$ eV (F4-TCNQ) and $\Delta_{h,HATCN}^{min} \approx 0.5$ eV (HATCN) is possible, which permits tailoring of the energy level alignment at these interfaces.

5.7.2. F4-TCNQ on ITO

Similar to the adsorption of F4-TCNQ on Ag(111) in Sec. 5.4, the deposition of F4-TCNQ on ITO increases the work function of the substrate as shown in the SECO spectra in Fig. 5.51a. For low nominal coverages (1, 2, and 4 Å) the work function increases from 4.5 eV (pristine UV/ozone cleaned ITO) almost linearly to 5.05 eV (4 Å F4-TCNQ coverage). With further deposition the work function increase becomes less, but the work function still rises up to a nominal coverage of 50 Å. Here, an absolute value of $\Phi_{ITO,F4-TCNQ} = 5.30$ eV is measured, which corresponds to a relative work function change of +0.8 eV compared to the pristine UV/ozone cleaned ITO. The evolution of the absolute work function is plotted in Fig. 5.52 versus the nominal coverage θ (black squares). The VB spectra recorded during the sequential deposition of F4-TCNQ on ITO are shown in Fig. 5.51b and c. Here, only a weak attenuation of the substrate features is observed. Even for nominal coverages of 50 Å only a weak molecular feature can be detected between 5 and 7 eV BE in the full VB spectrum in Fig. 5.51b. However, the zoom into the near Fermi level region in Fig. 5.51c reveals another molecule derived feature between 0.5 to 2.5 eV BE. This feature is assigned to two substrate-molecule hybrid states, one derived from the HOMO (H') and the other from the LUMO (L') of the neutral molecule in agreement with Braun and coworkers [155], who have observed a similarly

Chapter 5. Results and Discussion

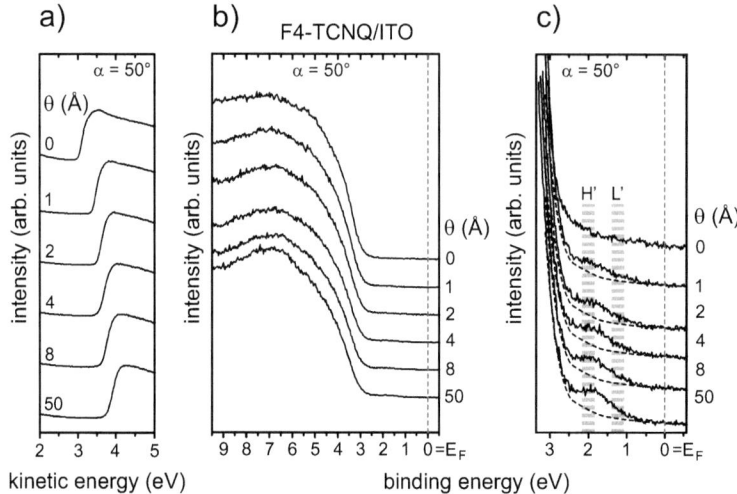

Figure 5.51.: The valence electronic structure of the sequential deposition of F4-TCNQ on ITO as obtained by UPS is shown. a) SECO spectra, b) full valence band spectra, and c) zoom into the near Fermi level region. H' and L' denote the substrate-molecule hybrid states derived from the HOMO (H') and LUMO (L') of neutral F4-TCNQ and E_F the Fermi level. The dashed black corresponds to the spectrum of pristine ITO and was used to obtain the area of the hybrid states, which is plotted in Fig. 5.52.

shaped feature after the adsorption of F4-TCNQ on AlO$_x$ substrates. The combined area of the states H' and L' is plotted in Fig. 5.52 (red circles). Its evolution with increasing nominal coverage θ is very similar to the evolution of the work function, which further supports that they originate from the CT. The steady increase in intensity of H' and L' leads to the conclusion, that even at 50 Å the first molecular layer is not fully closed. Furthermore, no attenuation of these features is observed, meaning that either high 3D islands build up or the multilayers are thermodynamically unstable and desorb. Strong islanding is also supported by the weak attenuation of the substrate's photoemission features. Multilayer desorption would stand in contrast to the adsorption of F4-TCNQ on Ag(111), where clearly multilayers were observable. However, Braun and coworkers do not explicitly report that multilayers are unstable at room temperature, but held their substrate at -100 °C during evaporation and measurement indicating some stability issues.

5.7. Tuning the hole injection barrier from indium tin oxide into α-NPD

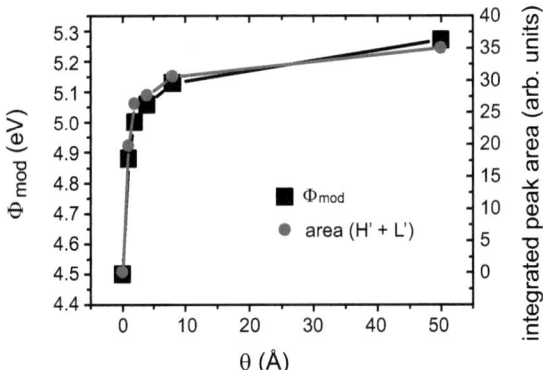

Figure 5.52.: Evolution of the absolute work function Φ_{mod} and the integrated peak area of the hybrid states H' and L' shown in Fig. 5.51 as a function of nominal F4-TCNQ coverage θ on ITO.

5.7.3. HATCN on ITO

The VB and SECO spectra obtained during the sequential deposition of HATCN on ITO are plotted in Fig. 5.53. The work function increases from 4.15 eV (pristine solution cleaned ITO) to 5.40 eV (nominally 100 Å HATCN) as shown in Fig. 5.53a. The evolution of the absolute work function versus the coverage is plotted in Fig. 5.54. Here, a linear increase of Φ for coverages up to 8 Å can be seen. For higher coverages, the work function increase becomes less and the work function only weakly rises with coverage, similar to the adsorption of F4-TCNQ on ITO. In contrast, the VB spectra in Fig. 5.53b and c show a good attenuation of the substrate's photoemission features, clearly indicating that HATCN multilayers can be grown on ITO. However, the attenuation is not as strong as for HATCN on Ag(111) and Cu(111), because of the higher roughness of the ITO. Thus, the ITO photoemission features are suppressed for coverages beyond 32 Å (compared to 19 Å for the metal substrates). The HOMO of the bulk is located at 5.0 eV BE having its onset at 4.15 eV BE, which leads to a bulk ionization energy of $IE_{ITO,HATCN} = 9.55$ eV, which is higher compared to HATCN in bulk films on coinage metal substrates. Despite the strong work function increase for the adsorption of HATCN on ITO, no emission from substrate-molecule hybrid states is observed in the valence band zoom in Fig. 5.53c. An explanation might be different adsorption sites, leading to different amounts of charge transfer and consequently to a very broad emission from these states compared to well defined states on ordered metal substrates. This is supported by

135

Chapter 5. Results and Discussion

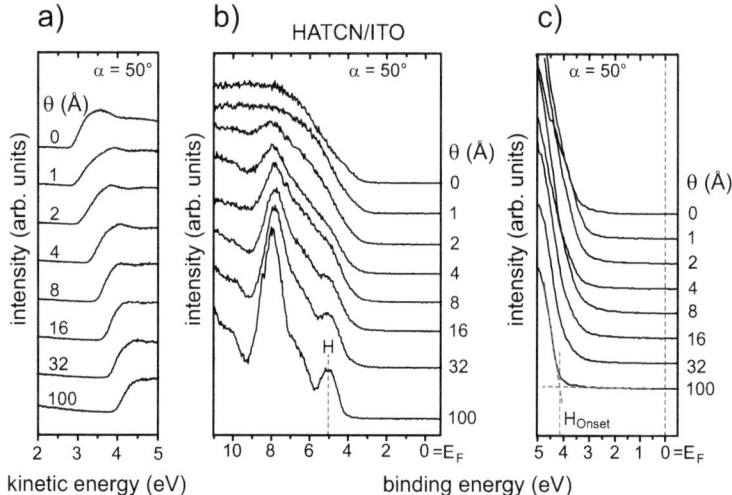

Figure 5.53.: The valence electronic structure of the sequential deposition of HATCN on ITO as obtained by UPS is shown. a) SECO spectra, b) full valence band spectra, and c) zoom into the near Fermi level region. H denotes the HOMO of the bulk film, H_{Onset} its onset and E_F the Fermi level.

the comparison of the hybrid states for the adsorption of F4-TCNQ on Ag(111) (Fig. 5.31b) and ITO (Fig. 5.51c), which are much more defined on the metal substrate, too.

5.7.4. Hole injection barriers from ITO into α−NPD

The strong work function modifications of the ITO substrates induced by the adsorption of F4-TCNQ and HATCN have been used to lower the hole injection barrier from ITO into α−NPD. The valence electronic structure for the adsorption of α−NPD on pristine solution cleaned ITO are shown in Fig. A.13 of the appendix. This yields a hole injection barrier of $\Delta_h = 1.25$ eV. Modifying the ITO substrate with different sub-monolayer coverages of F4-TCNQ and HATCN leads to a rigid shift of all molecular orbitals to lower BE, which effectively reduces the Δ_h. In Fig. 5.55 Δ_h from ITO into α−NPD is plotted as a function of Φ_{mod} (the F4-TCNQ/HATCN modified work function of the substrate). Two regimes are found for both electron accepting molecular materials: up to a certain critical Φ_{mod}^{pin} a linear relationship between the work function increase and the decrease in Δ_h is found (regime (i)). For higher work functions in regime (ii), Δ_h remains constant as the HOMO of α−NPD gets pinned at the Fermi level as already observed for other materials [49, 204].

5.7. Tuning the hole injection barrier from indium tin oxide into α-NPD

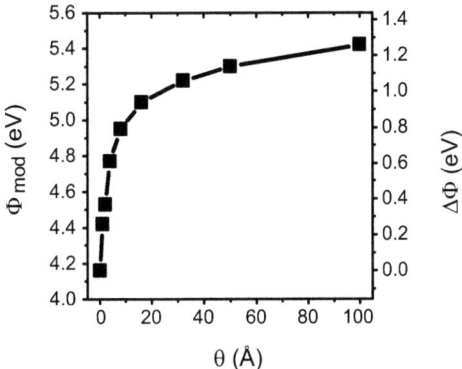

Figure 5.54.: Evolution of the absolute work function Φ_{mod} as a function of nominal HATCN coverage θ on ITO.

In regime (i) the Fermi level of the substrate lies in the HOMO-LUMO gap of α−NPD and thus vacuum level alignment occurs, resulting in a linear decrease of the hole injection barrier. For the start of regime (ii), different Φ_{mod}^{pin} are found depending on the electron acceptor used for the work function increase as shown in Fig. 5.55. In the case of F4-TCNQ on ITO, the critical work function lies around $\Phi_{mod,F4-TCNQ}^{pin} = 4.50$ eV. For higher work functions Δ_h remains constant at $\Delta_{h,F4-TCNQ}^{min} = 0.90$ eV. In contrast, Δ_h could be reduced even further to $\Delta_{h,HATCN}^{min} = 0.50$ eV by using HATCN. Here, the critical work function was found to be $\Phi_{mod,HATCN}^{pin} = 4.90$ eV. Since the pinning level should be an intrinsic materials property of α−NPD, the interfaces between the two electron accepting materials (F4-TNCQ and HATCN) and α−NPD were analyzed in more detail to find the reason for the different pinning levels. Fig. 5.56 shows the region of the CN stretch mode in the transmission infrared spectra of co-deposited F4-TCNQ/α−NPD (a) and HATCN/α−NPD (b) films on silicon oxide. As a reference for neutral F4-TCNQ, the bulk spectrum obtained on Ag(111) and for neutral HATCN the spectrum of the HATCN powder pressed in KBr is plotted. Fig. 5.56a shows that the symmetric stretching vibration of F4-TCNQ in the co-evaporated film (red line) is shifted by 9 cm⁻¹ to lower wavenumbers compared to the neutral F4-TCNQ film. This indicates a CT type reaction between F4-TCNQ and α−NPD, in which the former becomes negatively and the latter positively charged. The CT is further supported by the low energetic difference between the acceptor LUMO and the α−NPD HOMO. The LUMO of F4-TCNQ is located 5.25 eV below the vacuum level [14], while the HOMO of α−NPD is located 5.40 eV below the vacuum level [205]. Thus, the difference is only 0.15 eV, making a CT feasible. In contrast, the CN stretching mode of HATCN is not observed to shift for the co-deposited

Chapter 5. Results and Discussion

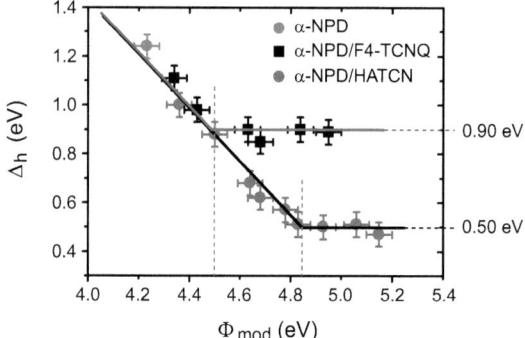

Figure 5.55.: Hole injection barrier Δ_h into nominally 30 Å thick α–NPD films as a function of F4-TCNQ/HATCN modified substrate work function Φ_{mod}. To adjust Φ_{mod} different pre-coverages of the acceptor materials have been used.

film (red line in Fig. 5.56b) compared to the neutral molecular spectrum. Furthermore, the energetic difference between the LUMO of HATCN and the HOMO of α–NPD is estimated to be 0.8 eV[6].

5.7.5. Conclusion

The strong electron acceptor molecules F4-TCNQ and HATCN have been deposited on ITO. The evolution of the electronic properties has been followed using UPS. In both cases the SECO spectra showed a strong work function increase as a function of nominal film thickness. For F4-TCNQ an absolute work function of 5.30 eV could be obtained, which is an increase of 0.80 eV compared to the pristine UV/ozone cleaned ITO substrate. The VB spectra showed two rather broad intensities in the energy gap of the neutral molecule, which could be attributed to substrate-molecule hybrid states. Their integrated intensity shows a similar behavior as the work function: a strong increase for coverages below ≈ 8 Å and a saturation for higher coverages. However, a slow increase is still observed at higher coverages, which is explained by tendency of F4-TCNQ to build 3D islands. This is also supported by the weak attenuation of the substrate's photoemission features in the VB. In case of the HATCN deposition on ITO, a maximum absolute work function of 5.40 eV has been measured, which is an increase of 1.25 eV compared to the pristine solution cleaned ITO. The work function changes its slope for coverages higher than 10 Å and reaches a saturation at higher coverages.

---

[6]No experimental data for the position of the LUMO of HATCN is available, but the difference in the calculated vertical EA of HATCN and F4-TCNQ is 0.6 eV. Consequently, the bulk LUMO position of HATCN is estimated to be 0.6 eV lower compared to F4-TCNQ.

5.7. Tuning the hole injection barrier from indium tin oxide into α-NPD

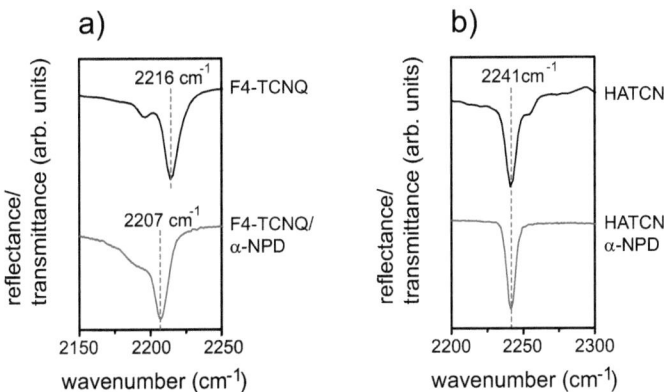

Figure 5.56.: Vibrational spectra showing the region of the CN stretching vibrations. a) thick F4-TCNQ film on Ag(111) (F4-TCNQ) and F4-TCNQ co-evaporated with α−NPD (F4-TCNQ/α−NPD). b) HATCN in KBr (HATCN) and HATCN co-evaporated with α−NPD (HATCN/α−NPD). The ratio of the co-evaporated films was in both cases 1:1. In all cases the symmetric CN stretching vibration is marked by the red dashed line and its position is given.

Even though this is an indication for strong CT between the substrate and the molecules, no new intensity in the energy gap of the neutral molecule was observed in the VB. The reason might be energetic disorder leading to a broadening of the respective hybrid orbitals or low photoemission cross-sections. Compared to the adsorption of F4-TCNQ on ITO, a stronger attenuation of the substrate's photoemission features is observed indicating a more compact layer growth.

The high work function F4-TCNQ/HATCN modified ITO substrates were further used as substrate for the deposition of an α−NPD film in order to test if a hole injection barrier lowering could be achieved. Δ_h for α−NPD on pristine ITO was found to be 1.25 eV. For the acceptor modified ITO substrates, Δ_h was decreased to a minimum value of 0.90 eV in the case of F4-TCNQ and even to 0.50 eV in the case of HATCN. Using different pre-coverages of the acceptor molecules and thus continuously changing the modified work function of the substrate, it was possible to decrease Δ_h linearly as a function of Φ_{mod}. A slope close to −1 was found, which is reminiscent of vacuum level alignment for the linear regime. Increasing the work function beyond 4.5 eV with F4-TCNQ did not result in a further lowering of Δ_h indicating that a pinning regime has been reached. In the case of HATCN the minimum Δ_h value was reached at 4.90 eV. Further increase of the work function did not result in a lowering of Δ_h. The results are reminiscent of a pinning of an α−NPD molecular level at

Chapter 5. Results and Discussion

the Fermi level. Since this should be an intrinsic materials property, the difference in the two levels must result from the different interaction between acceptors and the HTM. Using transmission infrared spectroscopy, it was shown that a CT type reaction occurs between F4-TCNQ and α−NPD, which did not occur in the case of HATCN. This was explained by the close vicinity of the α−NPD HOMO and the F4-TCNQ LUMO, which is energetically deeper lying compared to the LUMO of HATCN. HATCN can thus be regarded as an interesting material in actual devices employing ITO as transparent anode material to improve/tailor the energy level alignment between substrate and HTM.

Chapter 6.
Summary and Outlook

The aim of the present work was to find and characterize new and strong electron donor and acceptor materials that allow for a (continuous) tuning of the energy level alignment at electrode-organic interfaces. An additional task was to find conjugated donor benchmark materials, because at the start of the present work no such materials, with which a reduction of the EIB had been shown, were available. Since it is not possible to judge a priori if a certain molecule will behave as a strong electron donor or acceptor on electrode surfaces an initial screening process, using the work function modification as selection criterion, was applied. The selected molecules were then subjected to an extensive complementary multi-technique characterization, to obtain insights into the interplay between morphology and electronic structure right at the electrode-COM interface. The frontier electronic levels were analyzed using UPS. Using XPS, chemical shifts were observed that allowed for the identification of chemical reactions at the investigated interfaces. Furthermore, information on the film growth was obtained from the attenuation of the substrate's photoemission features and auger transitions using PES and AES. The morphology and molecular orientation during *in-situ* growth was monitored using RAIRS. Additional information of the molecular orientation was drawn from the PES selection rules. The experimental work was supported by theoretical modeling using DFT, which were done by collaborators.

The initial molecular screening was performed with metal single crystal substrates. They were used because they provide different reactivities and work functions. One clear trend was revealed during the experiments: The maximum positive (negative) work function modification increases (decreases) with decreasing substrate work function Φ_S. The spread in work function modification $\Delta\Phi$ is ca. 2.5 eV, meaning that the work function of the substrate metals can be tuned within this range by the molecules screened in this work. However, as mentioned before the modification range shifts depending on the substrate's work function Φ_S. Moreover, a clear correlation between calculated IE/EA and the donor/acceptor properties was experimentally confirmed. The strongest positive work function modifications were observed for F4-TCNQ and HATCN. Consequently, both were subjected to the in-depth analysis. On the other side, the by far strongest negative work function modifications were

Chapter 6. Summary and Outlook

observed for MV0 and thus investigated in more detail. With regard to MV0, NMA showed a weaker work function reduction potential. However, it was also chosen for a more detailed analysis, because of its strong similarities in the chemical structure and larger molecular weight compared to MV0.

The adsorption of MV0 was observed to occur from the vapor phase, which stresses that the evaporation temperature must be close to room temperature. In addition, multilayer coverages were not observed in the experiments. The occupied orbitals of a saturated monolayer of MV0 were found to be pinned with respect to the Fermi level on all three metal substrates. This was indicated by same position of deeper lying orbitals and the same final work function of the modified surfaces of $\Phi_{mod} = 3.3$ eV. The work function reductions were 2.20 eV on Au(111), 1.55 eV on Cu(111), and 1.20 eV on Ag(111), which underlined the strong electron donation potential of MV0. Theoretical modeling shows that the bonding pattern of MV0 changes upon charge transfer from quinoid to benzoid. This stabilizes the charge on the molecule and can be sought of another driving force for the electron transfer besides the Fermi level pinning. Similar valence and core electronic structures were found for MV0 on silver and copper surfaces, while they were distinctively different on gold. The orientation of MV0 on silver and copper in the saturated monolayer was observed to be face-on using RAIRS. The result is supported by PES selection rule considerations for the angular intensity dependence of the HOMO in the VB spectra. Unfortunately, the exact orientation of MV0 on gold remains unresolved within this work. However, core level analysis shows strong indications that the molecules are not oriented face-on. Further studies need to be conducted in order to shed light on this issue. The low work function surface of MV0 modified Au(111) and Ag(111) were further used to realize low electron injection barriers into subsequently deposited organic electron transport materials. The reduction of the EIB from gold into C_{60} was 0.65 eV, while it was 0.80 eV into Alq$_3$. On Ag(111) the reduction was 0.50 eV for C_{60} and 0.10 eV for Alq$_3$. The barrier reductions can be explained by different energy level alignment mechanisms for C_{60} and Alq$_3$: Due to the smaller transport gap of C_{60} Fermi level pinning occurs, whereas for Alq$_3$ vacuum level alignment occurs on the MV0 modified substrates. MV0 has proven to be a strong molecular donor, with a potential to reduce Δ_e into subsequently deposited electron transport materials and shown that the concept of injection barrier lowering also works at the cathode side of devices.

In an attempt to overcome the intrinsic limitations of MV0, NMA was synthesized by collaborators, which presents molecular donor with a larger molecular weight and structural similarities to MV0. Its electronic and structural properties in contact with metal substrates were analyzed in detail using PES. A linear reduction of the work function was found on all three substrates for submonolayer coverages. The maximum reduction values were 1.40 eV on Au(111), 1.15 eV on Cu(111), and 1.00 eV on Ag(111). They were reached at the monolayer coverage and are on all three coinage metal substrates beyond values reported for pure "push-back". Additional indication for a weak charge donation of NMA to the substrate are derived from differential shifts in the core level peak positions. The growth of NMA was observed to be layer-by-layer like. In analogy to the approach applied using MV0, the reduction of electron injection barriers were demonstrated using the prototypical electron transport material Alq$_3$. Thereby it was found that a linear tuning of Δ_e depending on the NMA pre-coverage up to the maximum reduction value of 0.25 eV was possible. The linear tuning of Δ_e presents another principal experimental proof done within this work, which extends the present knowledge on the modification of anodes (tuning of Δ_h using F4-TCNQ) to the cathode side of devices. Larger molecular structures, that are based on MV0, thus present a promising way for finding strong and more production process adequate donor materials that can be used to tune electron injection barriers.

Within this work, the strongest work function increases have been demonstrated for molecules adsorbed on silver surfaces. Since no experimental results were initially available for F4-TCNQ on this surface, a thorough characterization has been conducted. In the concourse of the experiments, the results were shown to be similar to the reported ones for F4-TCNQ on gold and copper surfaces: A strong and linear work function increase up to a maximum value of $\Phi_{mod} = 5.15$ eV coupled with a filling of a metal-molecule hybrid state derived from the former LUMO of the neutral molecule was observed. In contrast to the adsorption on gold and copper, the amount of charge transfer was found to change during monolayer formation. However, the application potential of F4-TCNQ is rather limited since diffusion through organic layers has been reported to occur at room temperature. Nevertheless, F4-TCNQ can serve as a benchmark acceptor system on all three coinage metal surfaces.

As a strong and large molecular weight acceptor, HATCN was chosen for the in-dept investigation. The maximum work function change was 1.00 eV on silver and 0.60 eV on copper and thus larger compared to F4-TCNQ. However, the evolution of the work function change as a function of coverage exhibited an additional regime at low coverages in which Φ_{mod} remained constant (silver) or even decreased (copper). With increasing molecular deposition, a linear increase of Φ_{mod} up to a saturation value (maximum) occurred. A phenomenon like this has not been reported before and could not be explained based on standard growth models. TDS and RAIRS experiments carried out within this work revealed that a re-orientation of initially face-on to edge-on HATCN molecules occurred. This was triggered by the molecular density present at the surface and facilitated through specific interactions of the peripheral molecular cyano-groups with the substrates. After all face-on molecules have been incorporated into an edge-on first molecular layer, further film formation proceeds mainly via upright oriented molecules. On gold the work function decreased initially by 0.50 eV, which is less compared to values reported for pure "push-back". Similar to Ag(111) and Cu(111), face-on molecules

Chapter 6. Summary and Outlook

were also detected in the initial stages of film growth here, but up to a coverage equivalent to *two* molecular layers. With increasing molecular density at the surface, edge-on oriented molecules become present and part of the face-on molecules re-orient. The growth is however more complex compared to copper and silver. In addition, the high work function electrode formed by HATCN on silver can be used to reduce the hole injection barrier into subsequently deposited α–NPD by 1 eV. This value is superior to Δ_h reductions for the same material using F4-TCNQ interlayers and proves the potential of HATCN.

The potential of HATCN to reduce the hole injection barriers on application relevant ITO substrates was presented in the last part of this work. The maximum work function increase of HATCN on ITO was 1.25 eV compared to 0.80 eV of the benchmark system F4-TCNQ on ITO. In both cases a strong and in part linear work function increase was found. The region of this linear increase was then used to continuously tune the hole injection barrier into subsequently deposited α–NPD. A linear relationship between Φ_{mod} and Δ_h was found until a pinning of the HOMO level of α–NPD occurred with respect to the Fermi level. However, the work function where the pinning was observed was higher for HATCN compared to F4-TCNQ. Consequently, the minimum hole injection barrier was 0.90 eV for F4-TCNQ and 0.50 eV for HATCN. This was explained by a CT reaction between F4-TCNQ and α–NPD, which did not occur in the case of HATCN.

Within this work, fundamental aspects of the interfaces between electrodes and conjugated organic materials present in all organic electronic devices have been addressed. On one hand, knowledge on molecular structures and their potential to modify the work function of electrode materials has been extended. Here, molecular electron donor materials emerged from initial molecule screening, which permitted to extend the concept of injection barrier tuning via strong charge transfer dipoles across an metal-organic interface to the cathode side of the device. Using this concept in future device architectures will allow to increase the current injection into the organic layers and widen the choice of electrode materials. Moreover, a large molecular weight acceptor emerged as potential substitute for F4-TCNQ and showed in most application relevant properties superior behavior. On the other hand, new fundamental insights into the properties of molecules adsorbed on metallic substrates were gained as a new interface phenomenon was revealed for the first time. The possible re-orientation of a strongly chemisorbed layer with increasing molecular density at the surface presents another fundamental mechanism that determines the electronic and morphological structure at metal-COM interfaces. This needs to be kept in mind for the design of future molecules and their application on clean metal substrates. All findings presented in the present work are valuable for the understanding of electrode-COM interfaces and will thus help in advancing the field of organic electronics.

Appendix A.

Appendix

A.1. Character and Correlation Tables

Character and correlation tables for the cases presented in this work. Tables are taken from Ref. [113].

Table A.1.: C_{2v} character table.

C_{2v}	I	C_2	σ_{xz}	σ_{yz}	
A_1	+1	+1	+1	+1	z
A_2	+1	+1	-1	-1	R_z
B_1	+1	-1	+1	-1	x, R_y
B_2	+1	-1	-1	+1	y, R_x

Table A.2.: C_{3v} character table.

C_{3v}	I	C	σ		
A_1	+1	+1	+1	+1	z
A_2	+1	+1	-1	-1	R_z
E	+2	-1	+1	0	x, y, R_y, R_y

Appendix A. Appendix

Table A.3.: D_{2h} character table.

$\mathbf{D_{2h}}$	I	$C_2(z)$	$C_2(y)$	$C_2(x)$	i	σ_{xy}	σ_{xz}	σ_{xz}	
A_g	+1	+1	+1	+1	+1	+1	+1	+1	x^2, y^2, z^2
B_{1g}	+1	+1	-1	-1	+1	+1	-1	-1	R_z, xy
B_{2g}	+1	-1	+1	-1	+1	-1	+1	-1	R_y, xz
B_{3g}	+1	-1	-1	+1	+1	-1	-1	+1	R_x, yz
A_u	+1	+1	+1	+1	-1	-1	-1	-1	xyz
B_{1u}	+1	+1	-1	-1	-1	-1	+1	+1	z, z^3, y^2z, xz^2
B_{2u}	+1	-1	+1	-1	-1	+1	-1	+1	y, yz^2, x^2y, y^3
B_{3u}	+1	-1	-1	+1	-1	+1	+1	-1	x, xz^2, xy^2z, x^3

Table A.4.: D_{3h} correlation table.

$\mathbf{D_{3h}}$	C_{3v}	$C_s(\sigma_{v,1})$	C_{2v}	$C_s(\sigma_h)$	$C_s(\sigma_{v,2})$
A_1'	A_1	A'	A_1	A'	A'
A_1''	A_2	A''	A_2	A''	A''
A_2'	A_2	A'	B_1	A'	A''
A_2''	A_1	A''	B_2	A''	A'
E'	E	A' + A''	$A_1 + B_1$	A'	A' + A''
E''	E	A' + A''	$A_2 + B_2$	A''	A' + A''

A.2. Image dipole theory and the surface selection rule

One can also arrive at the surface selection rule (Sec. 3.2.2) using image dipole theory as shown in Fig. A.1. The long-range electromagnetic field of the IR radiation cannot distinguish the dipole and its image and will therefore interact with the sum of the dipole fields. In Fig. A.1a, the dipole is oriented perpendicular to the surface leading to an increased absorption. In the case of a parallel dipole, as shown in Fig. A.1b, the sum of the dipole fields yields a quadrupolar field, which is not excitable by the IR radiation. In other words, the dipole is effectively screened by its own image.

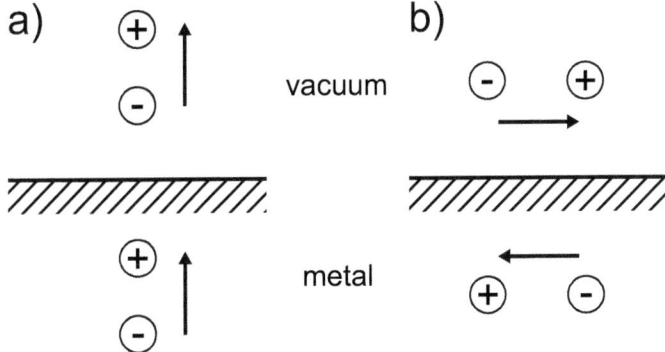

Figure A.1.: Metal-vacuum interface showing dipoles above the surface and their image in the metal. a) dipole perpendicular to the surface and b) parallel to the surface. The arrows indicate the direction of the dipole moment.

A.3. Beam energies and photon flux

Table A.5.: Experimentally used beam energies and approximate photon flux at the sample position. Additional values taken from [206, 91].

Synchrotron beamline	Energy (eV)	Width (meV)	Intensity (%)	Flux (photons/ sec·100 mA ring current)
BESSY II PM4	35	3.7	100	1×10^8
BESSY II PM4	620	270	100	5×10^9
HASYLAB E1	22		100	1×10^{10}

Line	Energy (eV)	Width (meV)	Intensity (%)	Flux (photons/sec)
He Iα	21.22	5	100	1×10^{12}
He Iβ	23.09		2	
He Iγ	23.74		0.5	
Mg K$\alpha_{1,2}$	1253.6	700	100	1×10^{12}
Mg Kα_3	1262.1		9	
Mg Kα_4	1263.7		5	
Al Kα_1	1486.6	850	100	1×10^{12}
Al Kα_1	1496.3		7	
Al Kα_1	1498.3		3	

A.4. C1s spectra of a saturated monolayer of MV0 on Ag(111), Cu(111), and Au(111)

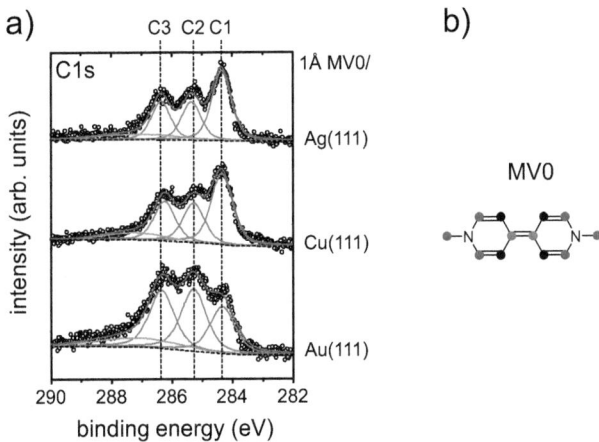

Figure A.2.: a) chemical structure of MV0, showing different carbon species according to Ref. [152]. b) carbon core level spectra of a saturated monolayer of MV0 on Ag(111), Cu(111), and Au(111). The spectral deconvolution has been done using mixed Gaussian and Lorentzian peaks and Shirley backgrounds.

The spectral deconvolution yields three distinct peaks located at 284.3 eV (C1), 285.3 eV (C2), and 286.3 eV BE (C3) with a FWHM of 0.8 eV on Ag(111). Their intensity ratio is approximately 2:1:1 (C1:C2:C3). Additionally a broad feature centered at 287.1 eV BE is found in for the adsorption of MV0 on all three substrates, which is attributed to shake-up processes. On copper, the C1 and C2 peaks are observed at the same positions, only the C3 peak is found at 0.3 eV lower BE (386.0 eV BE) compared to Ag(111). The FWHM is the same on copper as on silver. The spectral deconvolution of the C1s signal on Au(111) yields three peaks, which are located at the same positions as on Ag(111). Certainly, the intensity ratios are much different and observed to be approximately 5:6.5:6.5 (C1:C2:C3). Similar to the N1s core levels, the FWHM of the C1s peaks is increased to 1.0 eV on Au(111) compared to Ag(111) and Cu(111). In the gas phase molecule, three different carbon species should be observed: one, which stems from the carbons bound to nitrogen (red dots in Fig. 5.11a), one from the carbons bound to hydrogen (black dots) and another one that is due to the central C=C bond (blue dots). In the gas phase molecule their ratio should be 3:2:1 (red:black:blue). However, the peak ratios observed in the C1s spectra shown in Fig. 5.10b are clearly different.

149

Appendix A. Appendix

Thus, the deconvolution of the carbon 1s spectra of MV0 adsorbed on the three coinage metal surfaces must remain unresolved within the present work.

A.5. Additional experimental results for NMA in solution and adsorbed on Ag(111), Cu(111), and Au(111)

The electrochemical and optical properties of NMA in solution were investigated by means of cyclicvoltammetry and optical absorption spectroscopy. The obtained voltammograms for NMA in Fig. A.3 display a single reversible redox wave corresponding to the oxidation at both acridanic nitrogen atoms. Note that the higher current density obtained with the GC disc compared to the gold disc is most probably due to the higher molecular coverage on the gold electrode surface. In fact, the difference in potential between the oxidation and reduction peak is about 30 mV, proving the simultaneous bi-electronic character of the electrochemical reaction [207]. The oxidation potentials referred to the ferrocene (Fc/Fc+) redox couple are the same for the two electrode materials (i.e. -0.28 V as indicated by the dashed black line in Fig. A.3); when reported vs. the absolute vacuum scale [208, 209] the corresponding HOMO level of NMA is positioned at -4.9 eV. The difference to the calculated IE ($IE_{vert,NMA} = 5.45$ eV) stems at least in part from the polarization of the solvent in the CV measurements that is not accounted for in the calculations.

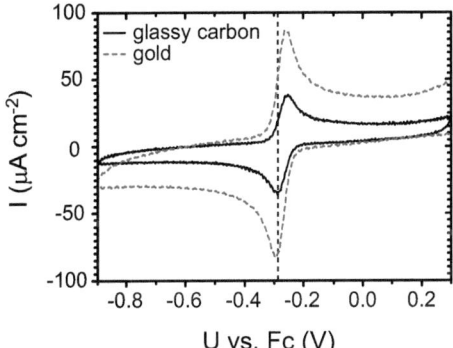

Figure A.3.: Voltammograms of NMA dissolved (concentration about 10^{-4} M) in a 0.1 M solution of tetrabutylammonium exafluorophosphate (TBAPF) and anhydrous acetonitrile (ACN) for two different working electrodes. The oxidation potentials reported here are referred to the Fc/Fc+ redox couple.

Fig. A.4a shows the optical absorption spectrum for a 10^{-5} M solution of NMA in its pure form and after oxidation with 0.1 μL of a 1 M SbCl$_5$ solution in the same solvent (dichloromethane). The first absorption maximum of neutral NMA peaks at 475 ± 2 nm (2.61 ± 0.05 eV) with the onset of the peak at 542 ± 2 nm, which corresponds to an optical band gap of (2.30 ± 0.05) eV in solution. After the double oxidation, the first absorption maximum

Appendix A. Appendix

is slightly shifted to the blue to 452 ± 2 nm (2.74 ± 0.05 eV) and the peak onset to 517 ± 2 nm (2.40 ± 0.05 eV); a second absorption maximum appears at 365 ± 2 nm (3.40 ± 0.05 eV).

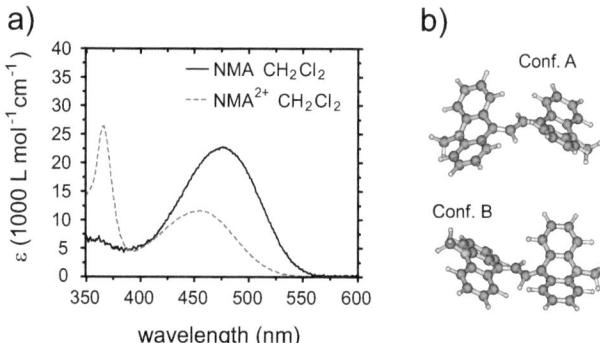

Figure A.4.: a) Molecular extinction coefficient (ϵ) as function of the wavelength for a 10^{-5} M solution of NMA in its pure form (black curve) and after solution oxidation with 0.1 μL of a 1 M SbCl$_5$ solution (dashed red curve) in the same solvent. b) The two (local) minimum configurations of the neutral molecule found using DFT-based as well as semiempirical calculations.

A.5.1. Full core level spectra of NMA on Au(111), Ag(111), and Cu(111)

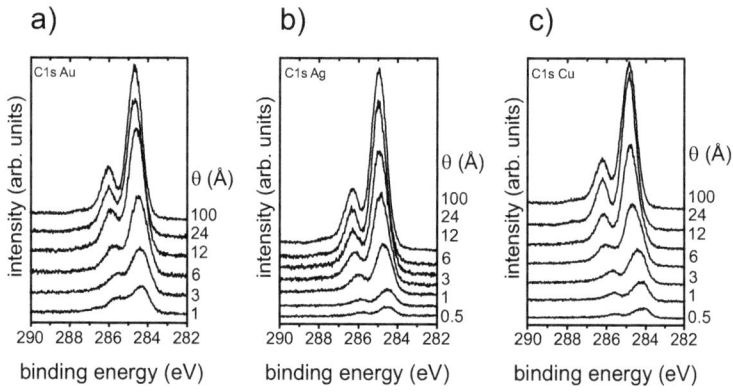

Figure A.5.: Carbon 1s core level spectra of NMA deposited sequentially on a) Au(111), b) Ag(111), and c) Cu(111).

152

A.5. Additional experimental results for NMA

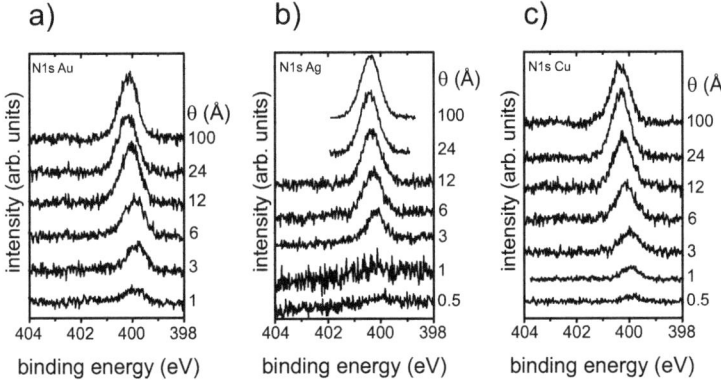

Figure A.6.: Nitrogen 1s core level spectra of NMA deposited sequentially on a) Au(111), b) Ag(111), and c) Cu(111).

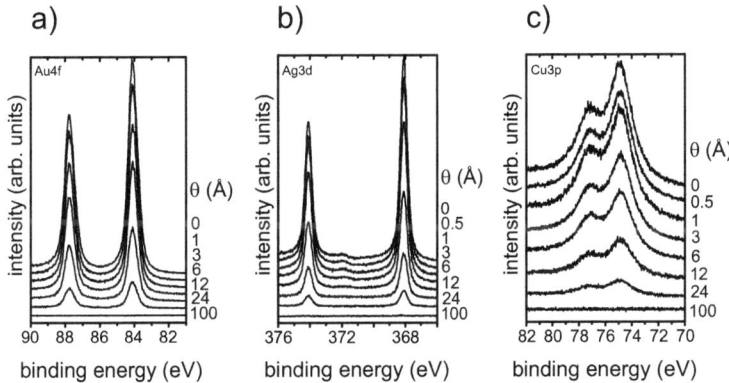

Figure A.7.: Substrate core level spectra of NMA deposited sequentially on a) Au(111) (Au 4f), b) Ag(111) (Ag 3d), and c) Cu(111) (Cu 3p).

Appendix A. Appendix

A.6. Vibrational modes of a thick F4-TCNQ film on Ag(111)

Wavenumber (cm^{-1})
802
878
896
908
973
1154
1203
1270
1351
1393
1496
1533
1568
1630
1676
1765
2196
2216

Table A.6.: Wavenumbers of all vibrational modes observed in the RAIR spectrum of a 20 Å thick F4-TCNQ film adsorbed on Ag(111) shown in Fig. 5.35.

A.7. Fitting of the RAIR spectra for thin films of HATCN on Ag(111) using a Fano-type line shape

The dotted spectrum in Fig. A.8 shows the experimental RAIRS spectrum of Fig. 5.39, which was collected from a HATCN film with a nominal thickness of 1.8 Å. The red line represents a fit to the experimental data using a Fano-type line shape from Ref. [197, 210]:

$$L(\nu) = c\gamma\nu\nu_r \frac{\left[1 - \frac{\tau}{\gamma}\left(\nu^2 - \nu_r^2\right)\right]^2}{(\gamma\nu)^2 + (\nu^2 - \nu_r^2)^2} \qquad (A.1)$$

with c being a constant used to adjust L to the experimental data, γ the full-width at half-maximum of the line shape, τ the tunneling rate, and ν_r the fully renormalized vibrational frequency.

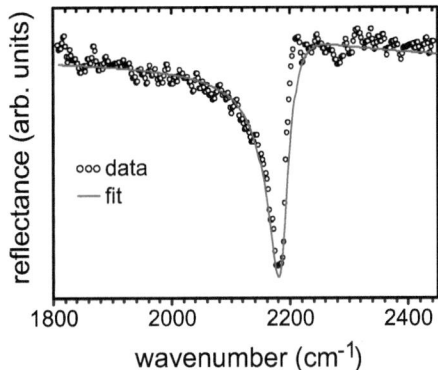

Figure A.8.: Fit using a Fano-type line shape to the data of the RAIRS spectrum of Fig. 5.39, which was collected from a HATCN film with a nominal thickness of 1.8 Å. The fitting parameters are $\gamma = 43$ cm^{-1}, $\tau = 0.000123$ cm and $r = 2185$ cm^{-1}.

Appendix A. Appendix

A.8. Theoretical description of the upright standing HATCN layer on Ag(111)

For the upright standing conformation, an orthogonal, $2 \times 3\sqrt{3}$ unit cell was assumed. This geometry corresponds to $\pi - \pi$-stacking distances between the individual HATCN cores of 5.88 Å and a lateral distance of 15.28 Å between the centers of neighboring molecules. Upright standing HATCN supposedly adsorbs with two CN groups pointing directly towards the surface. There are two different possibilities to realize this situation. Either, the molecule could adsorb on its "cusp", i.e. with adjacent cyano groups (Fig. A.9b). Alternatively, the molecule could adsorb 60° rotated to the former structure, and two cyano groups of different rings point toward the surface (Fig. A.9a). Electronically, both geometries yield similar results. The work function modification is computed to be +2.48 eV for the "cusp"-geometry and +2.37 eV for the 60° rotated one. Slight geometrical distortions can be observed for both conformations. These are attributed to the incommensurability between the CN group spacing and the spacing between the optimal docking site. In neither geometry is a molecular dipole induced ($\mu_{Mol} < 0.1$ eV). All work-function modifications must therefore be attributed to charge-transfer processes.

A.8. Theoretical description of the upright standing HATCN layer on Ag(111)

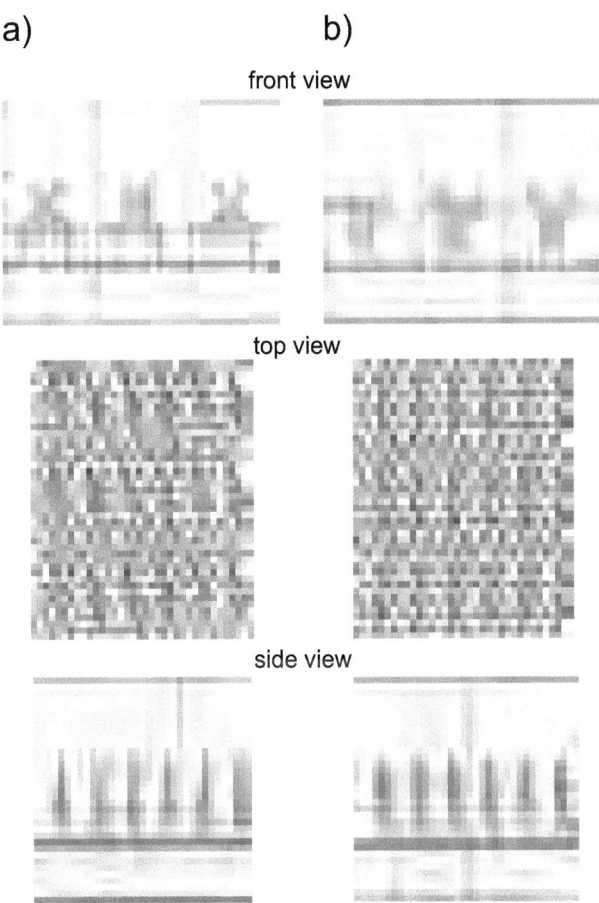

Figure A.9.: Front view (top) top view (middle) and side view (bottom) of upright standing HATCN adsorbing with its cusp (b) and 60° rotated (a). The unit cell is reproduced 6 times in the x and 3 times in the y direction.

Appendix A. Appendix

A.9. Auger electron spectroscopy

The differential Auger spectra taken during the sequential deposition of HATCN on Ag(111) for the region of the Ag MVV transition are shown in Fig. A.10a. The normalized peak to peak height of this transition is plotted in Fig. A.10b for the coverage values given in Fig. A.10a. Two regimes with different slopes are observed, with the break located at about 14 Å. This indicates a change of the growth mode at this coverage [211, 212], which is ascribed to the saturation of the first standing monolayer of HATCN on Ag(111). Beyond this coverage value the growth mode changes and thus the attenuation of the Ag substrate is different. The normalized Ag Auger signal for lower coverages of HATCN on Ag(111) with a smaller step size is shown in Fig. A.11. Here also a clear change in the slope is observed which occurs between 1 and 2 Å deposition, which agrees very well with the reported coverage values of the closed face-on monolayer.

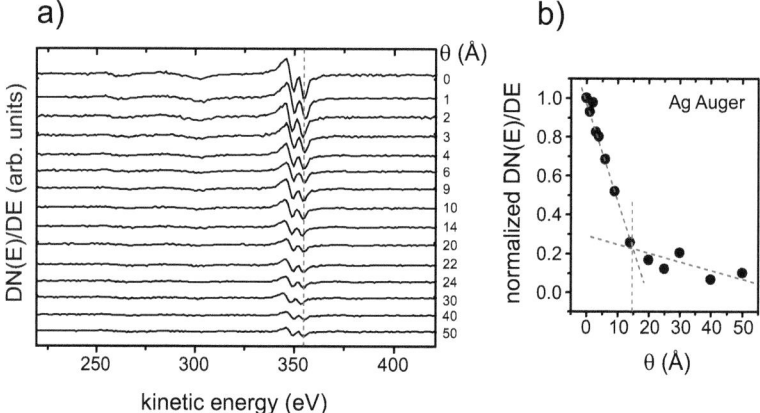

Figure A.10.: a) Differential Auger spectra of the Ag MVV transition for different coverages of HATCN on Ag(111). b) Normalized peak to peak intensity of the Ag MNV transition plotted versus the coverage (θ).

A.9. Auger electron spectroscopy

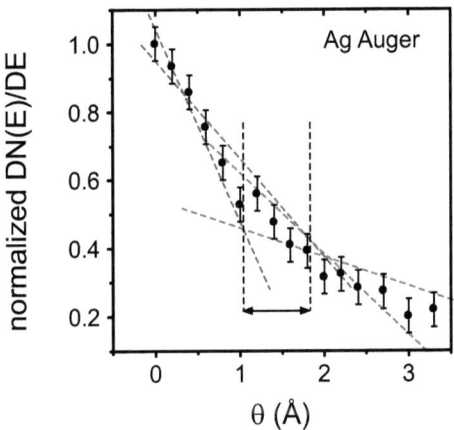

Figure A.11.: Normalized peak to peak intensity of the Ag MVV transition plotted versus the coverage (θ) (Low coverage regime).

Appendix A. Appendix

A.10. STM results for the adsorption of HATCN on Ag(111) and Cu(111)

The following STM images have been obtained by Hendrik Glowatzki (Humboldt-Universität zu Berlin).

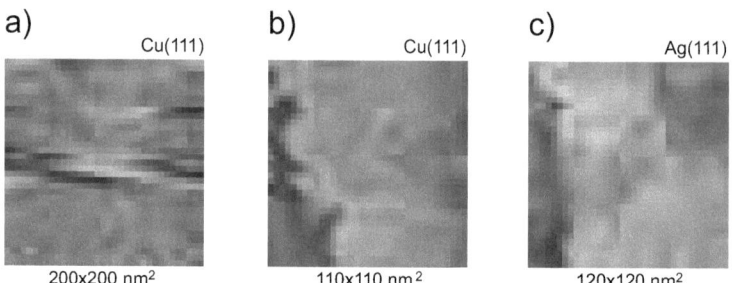

Figure A.12.: Constant current STM images of HATCN adsorbed on Cu(111) (a, b) and Ag(111) (c). The parameters are: a) ≈1 Å HATCN, $U = -1$ V, $I = 1$ nA; b) ≈3 Å HATCN, $U = -1$ V, $I = 0.3$ nA; c) ≈2 Å HATCN, $U = -1.2$ V, $I = 0.2$ nA.

A.11. Valence electronic structure of α−NPD on ITO

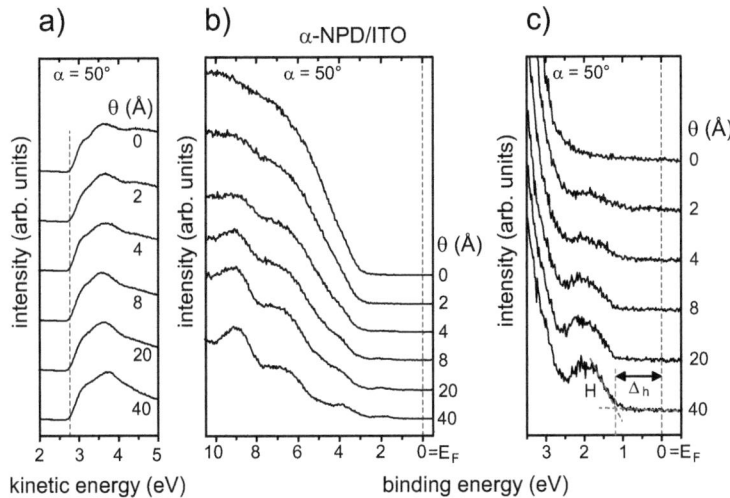

Figure A.13.: The valence electronic structure of the sequential deposition of α−NPD on ITO as obtained by UPS is shown. a) SECO spectra, b) full valence band spectra, and c) zoom into the near Fermi level region. H denotes the HOMO of the bulk film and E_F the Fermi level. The hole injection barrier for the 40 Å film is $\Delta_h = 1.25$ eV.

Bibliography

[1] POPE, M. ; SWENBERG, C.E.: *Electronic processes in organic crystals and polymers.* Oxford University Press, USA, 1999

[2] HEEGER, A.J.: Nobel Lecture: Semiconducting and metallic polymers: The fourth generation of polymeric materials. In: *Rev. Mod. Phys.* 73 (2001), Nr. 3, S. 681–700

[3] KARL, N.: Charge carrier transport in organic semiconductors. In: *Synth. Met.* 133 (2003), S. 649–657

[4] TANG, C. W. ; VANSLYKE, S. A.: Organic electroluminescent diodes. In: *Appl. Phys. Lett.* 51 (1987), S. 913–915

[5] MALLIARAS, G. ; FRIEND, R.: An Organic Electronics Primer. In: *Physics Today* 58 (2005), Nr. 5, S. 53. – ISSN 00319228

[6] ANTHONY, J.E.: Functionalized Acenes and Heteroacenes for Organic Electronics. In: *Chem. Rev.* 106 (2006), Nr. 12, S. 5028–5048

[7] BALDO, M. A. ; FORREST, S. R.: Interface-limited injection in amorphous organic semiconductors. In: *Phys. Rev. B* 64 (2001), S. 085201

[8] DI, C. ; YU, G. ; LIU, Y. ; XU, X. ; SONG, Y. ; ZHU, D.: Effective modification of indium tin oxide for improved hole injection in organic light-emitting devices. In: *Appl. Phys. Lett.* 89 (2006), S. 033502

[9] KOCH, N. ; DUHM, S. ; RABE, J. P. ; VOLLMER, A. ; JOHNSON, R. L.: Optimized Hole Injection with Strong Electron Acceptors at Organic-Metal Interfaces. In: *Phys. Rev. Lett.* 95 (2005), S. 237601

[10] OSIKOWICZ, W. ; CRISPIN, X. ; TENGSTEDT, C. ; LINDELL, L. ; KUGLER, T. ; SALANECK, W. R.: Transparent low-work-function indium tin oxide electrode obtained by molecular scale interface engineering. In: *Appl. Phys. Lett.* 85 (2004), Nr. 9, S. 1616–1618

Bibliography

[11] LINDELL, L. ; UNGE, M. ; OSIKOWICZ, W. ; STAFSTRÖM, S. ; W.R.SALANECK ; CRISPIN, X. ; JONG, M.P. de: Integer charge transfer at the tetrakis (dimethylamino) ethylene/Au interface. In: *Appl. Phys. Lett.* 92 (2008), S. 163302

[12] CHAN, C. K. ; KAHN, A. ; ZHANG, Q. ; BARLOW, S. ; MARDER, S. R.: Incorporation of cobaltocene as an n-dopant in organic molecular films. In: *J. Appl. Phys.* 102 (2007), S. 014906

[13] CHAN, C. K. ; ZHAO, W. ; BARLOW, S. ; MARDER, S. ; KAHN, A.: Decamethylcobaltocene as an efficient n-dopant in organic electronic materials and devices. In: *Org. Electron.* 9 (2008), Nr. 5, S. 575–581. – ISSN 1566–1199

[14] GAO, W. ; KAHN, A.: Controlled p-doping of zinc phthalocyanine by coevaporation with tetrafluorotetracyanoquinodimethane: A direct and inverse photoemission study. In: *Appl. Phys. Lett.* 79 (2001), S. 4040–4042

[15] DUHM, S. ; SALZMANN, I. ; BRÖKER, B. ; GLOWATZKI, H. ; JOHNSON, R. L. ; KOCH, N.: Interdiffusion of molecular acceptors through organic layers to metal substrates mimics doping-related energy level shifts. In: *Appl. Phys. Lett.* 95 (2009), Nr. 9, S. 093305–3

[16] ROMANER, L. ; HEIMEL, G. ; BRÉDAS, J.-L. ; GERLACH, A. ; SCHREIBER, F. ; JOHNSON, R. L. ; ZEGENHAGEN, J. ; DUHM, S. ; KOCH, N. ; ZOJER, E.: Impact of Bidirectional Charge Transfer and Molecular Distortions on the Electronic Structure of a Metal-Organic Interface. In: *Phys. Rev. Lett.* 99 (2007), S. 256801

[17] KOCH, N.: Organic Electronic Devices and Their Functional Interfaces. In: *Chem. Phys. Chem.* 8 (2007), Nr. 10, S. 1438–1455

[18] LANG, N. D. ; KOHN, W: Theory of Metal Surfaces: Work Function. In: *Phys. Rev. B* 3 (1971), S. 1215–1223

[19] LANG, N. D. ; KOHN, W.: Theory of Metal Surfaces: Charge Density and Surface Energy. In: *Phys. Rev. B* 1 (1970), Nr. 12, S. 4555

[20] CRISPIN, X. ; GESKIN, V. ; CRISPIN, A. ; CORNIL, J. ; LAZZARONI, R. ; SALANECK, W. R. ; BREDAS, J.-L.: Characterization of the Interface Dipole at Organic/ Metal Interfaces. In: *J. Am. Chem. Soc.* 124 (2002), S. 8131–8141

[21] SCHWOERER, M. ; WOLF, H. C.: *Organic Molecular Solids.* Wiley-VCH, 2006

[22] SILINSH, E. A.: *Organic Molecular Crystals.* Springer, Berlin, 1980

Bibliography

[23] HILL, I. G. ; KAHN, A. ; CORNIL, J. ; SANTOS, D. A. ; BRÉDAS, J. L.: Occupied and unoccupied electronic levels in organic [pi]-conjugated molecules: comparison between experiment and theory. In: *Chem. Phys. Lett.* 317 (2000), Nr. 3-5, S. 444–450

[24] HILL, I. G. ; KAHN, A. ; SOOS, Z. G. ; PASCAL, JR., R. A.: Charge-separation energy in films of π-conjugated organic molecules. In: *Chem. Phys. Lett.* 327 (2000), S. 181–188

[25] ZAHN, D.R.T. ; GAVRILA, G. N. ; GORGOI, M.: The transport gap of organic semiconductors studied using the combination of direct and inverse photoemission. In: *Chem. Phys.* 325 (2006), Nr. 1, S. 99–112. – ISSN 0301–0104

[26] KRAUSE, S. ; CASU, M. B. ; SCHÖLL, A. ; UMBACH, E.: Determination of transport levels of organic semiconductors by UPS and IPS. In: *New Journal of Physics* 10 (2008), Nr. 085001, S. 085001

[27] GREGG, B. A. ; HANNA, M. C.: Comparing organic to inorganic photovoltaic cells: Theory, experiment, and simulation. In: *Journal of Applied Physics* 93 (2003), Nr. 6, S. 3605. – ISSN 00218979

[28] SMOLUCHOWSKI, R.: Anisotropy of the Electronic Work Function of Metals. In: *Phys. Rev.* 60 (1941), S. 661–674

[29] NATAN, A. ; KRONIK, L. ; HAICK, H. ; TUNG, R.T.: Electrostatic Properties of Ideal and Non-ideal Polar Organic Monolayers: Implications for Electronic Devices. In: *Adv. Mater.* 19 (2007), Nr. 23, S. 4103–4117

[30] DUHM, S. ; HEIMEL, G. ; SALZMANN, I. ; GLOWATZKI, H. ; JOHNSON, R. L. ; VOLLMER, A. ; RABE, J. P. ; KOCH, N.: Orientation-dependent ionization energies and interface dipoles in ordered molecular assemblies. In: *Nature Mater.* 7 (2008), S. 326–332

[31] SALZMANN, I. ; DUHM, S. ; HEIMEL, G. ; OEHZELT, M. ; KNIPRATH, R. ; JOHNSON, R. L. ; RABE, J. P. ; KOCH, N.: Tuning the Ionization Energy of Organic Semiconductor Films: The Role of Intramolecular Polar Bonds. In: *J. Am. Chem. Soc.* 130 (2008), S. 12870–12871

[32] HAEKKINEN, H. ; YOON, B. ; LANDMAN, U. ; LI, X. ; ZHAI, H.-J. ; WANG, L.-S.: On the Electronic and Atomic Structures of Small Au Clusters: A Photoelectron Spectroscopy and Density-Functional Study. In: *ChemInform* 34 (2003), Nr. 45

[33] LI, J. ; LI, X. ; ZHAI, H. J. ; WANG, L. S.: Au20: A Tetrahedral Cluster. In: *Science* 299 (2003), S. 864–867

[34] ATKINS, P.W.: *Physical Chemistry*. W.H. Freeman & Company, 1982

Bibliography

[35] KOCH, N. ; GERLACH, A. ; DUHM, S. ; GLOWATZKI, H. ; HEIMEL, G. ; VOLLMER, A. ; SAKAMOTO, Y. ; SUZUKI, T. ; ZEGENHAGEN, J. ; RABE, J. P. ; SCHREIBER, F.: Adsorption-Induced Intramolecular Dipole: Correlating Molecular Conformation and Interface Electronic Structure. In: *J. Am. Chem. Soc.* 130 (2008), S. 7300–7304

[36] DUHM, S. ; GERLACH, A. ; SALZMANN, I. ; BRÖKER, B. ; JOHNSON, R. L. ; SCHREIBER, F. ; KOCH, N.: PTCDA on Au(111), Ag(111) and Cu(111): Correlation of interface charge transfer to bonding distance. In: *Org. Electron.* 9 (2008), S. 111–118

[37] KAHN, A. ; KOCH, N. ; GAO, W. Y.: Electronic structure and electrical properties of interfaces between metals and π-conjugated molecular films. In: *J. Poly. Sci. B* 41 (2003), S. 2529–2548

[38] ISHII, H. ; SUGIYAMA, K. ; ITO, E. ; SEKI, K.: Energy Level Alignment and Interfacial Electronic Structures at Organic/Metal and Organic/Organic Interfaces. In: *Adv. Mater.* 11 (1999), S. 605–625

[39] LANG, N. D.: Interaction between Closed-Shell Systems and Metal Surfaces. In: *Physical Review Letters* 46 (1981), Nr. 13, S. 842

[40] BAGUS, P. S. ; KÄFER, D. ; WITTE, G. ; WÖLL, C.: Work Function Changes Induced by Charged Adsorbates: Origin of the Polarity Asymmetry. In: *Phys. Rev. Lett.* 100 (2008), Nr. 12, S. 126101–4

[41] DUDDE, R. ; REIHL, B.: Complete electronic structure of oriented films of hexatriacontane. In: *Chem. Phys. Lett.* 196 (1992), Nr. 1-2, S. 91–96. – ISSN 0009–2614

[42] YOSHIMURA, D. ; ISHII, H. ; OUCHI, Y. ; ITO, E. ; MIYAMAE, T. ; HASEGAWA, S. ; OKUDAIRA, K. K. ; UENO, N. ; SEKI, K.: Angle-resolved ultraviolet photoelectron spectroscopy and theoretical simulation of a well-ordered ultrathin film of tetratetracontane (n-C44H90) on Cu(100): Molecular orientation and intramolecular energy-band dispersion. In: *Phys. Rev. B* 60 (1999), Nr. 12, S. 9046

[43] ITO, E. ; OJI, H. ; ISHII, H. ; OICHI, K. ; OUCHI, Y. ; SEKI, K.: Interfacial electronic structure of long-chain alkane/metal systems studied by UV-photoelectron and metastable atom electron spectroscopies. In: *Chem. Phys. Lett.* 287 (1998), S. 137–142

[44] WAN, A. ; HWANG, J. ; AMY, F. ; KAHN, A.: Impact of electrode contamination on the [alpha]-NPD/Au hole injection barrier. In: *Org. Electron.* 6 (2005), Nr. 1, S. 47–54

[45] GROBOSCH, M. ; KNUPFER, M.: Charge-Injection Barriers at Realistic Metal/Organic Interfaces: Metals Become Faceless. In: *Adv. Mater.* 19 (2007), Nr. 5, S. 754–756

Bibliography

[46] BAGUS, P. S. ; STAEMMLER, V. ; WÖLL, C.: Exchangelike Effects for Closed-Shell Adsorbates: Interface Dipole and Work Function. In: *Phys. Rev. Lett.* 89 (2002), S. 096104

[47] HWANG, J. ; WAN, A. ; KAHN, A.: Energetics of metal-organic interfaces: New experiments and assessment of the field. In: *Mater Sci Eng R* 64 (2009), Nr. 1-2, S. 1–31. – ISSN 0927-796X

[48] FAHLMAN, M ; CRISPIN, A ; CRISPIN, X ; HENZE, S K M. ; JONG, M P. ; OSIKOWICZ, W ; TENGSTEDT, C ; SALANECK, W R.: Electronic structure of hybrid interfaces for polymer-based electronics. In: *J Phys : Condens Matter* 19 (2007), S. 183202

[49] BRAUN, S. ; SALANECK, W. R. ; FAHLMAN, M.: Energy-Level Alignment at Organic/Metal and Organic/Organic Interfaces. In: *Adv. Mater.* 21 (2009), Nr. 14-15, S. 1450–1472

[50] ZOU, Y. ; KILIAN, L. ; SCHÖLL, A. ; SCHMIDT, Th. ; FINK, R. ; UMBACH, E.: Chemical bonding of PTCDA on Ag surfaces and the formation of interface states. In: *Surf. Sci.* 600 (2006), S. 1240–1251

[51] SCOTT, J.C. ; MALLIARAS, G.G.: Charge injection and recombination at the metal-organic interface. In: *Chem. Phys. Lett.* 299 (1999), Nr. 2, S. 115–119

[52] SZE, S. M.: *Physics of Semiconductor Devices*. John Wiley and Sons, 1981

[53] VENABLES, J. A. ; SPILLER, G. D. T. ; HANBUCKEN, M.: Nucleation and growth of thin films. In: *Reports on Progress in Physics* 47 (1984), Nr. 4, S. 399–459. – ISSN 0034-4885

[54] LUKAS, S. ; WITTE, G. ; WÖLL, Ch.: Novel Mechanism for Molecular Self-Assembly on Metal Substrates: Unidirectional Rows of Pentacene on Cu(110) Produced by a Substrate-Mediated Repulsion. In: *Phys. Rev. Lett.* 88 (2001), Nr. 2, S. 028301. – ISSN 0031-9007

[55] HA, S. D. ; KAAFARANI, B. R. ; BARLOW, S. ; MARDER, S. R. ; KAHN, A.: Multiphase Growth and Electronic Structure of Ultrathin Hexaazatrinaphthylene on Au(111). In: *J. Phys. Chem. C* 111 (2007), Nr. 28, S. 10493–10497

[56] BARABÁSI, A. L. ; STANLEY, H. E.: *Fractal concepts in surface growth*. Cambridge University Press, 1995

[57] PIMPINELLI, A. ; VILLAIN, J.: *Physics of crystal growth*. Cambridge University Press, 1998

Bibliography

[58] SCHREIBER, F.: Organic molecular beam deposition: Growth studies beyond the first monolayer. In: *Phys Status Solidi A* 201 (2004), Nr. 6, S. 1037–1054

[59] BISCARINI, F. ; SAMORI, P. ; GRECO, O. ; ZAMBONI, R.: Scaling Behavior of Anisotropic Organic Thin Films Grown in High Vacuum. In: *Phys. Rev. Lett.* 78 (1997), Nr. 12, S. 2389–2392. – ISSN 0031–9007

[60] DÜRR, A. C. ; SCHREIBER, F. ; RITLEY, K. A. ; KRUPPA, V. ; KRUG, J. ; DOSCH, H. ; STRUTH, B.: Rapid Roughening in Thin Film Growth of an Organic Semiconductor (Diindenoperylene). In: *Phys. Rev. Lett.* 90 (2003), Nr. 1, S. 016104

[61] WITTE, G. ; HÄNEL, K. ; SÖHNCHEN, S. ; WÖLL, Ch.: Growth and morphology of thin films of aromatic molecules on metals: the case of perylene. In: *Applied Physics A: Materials Science & Processing* 82 (2006), Nr. 3, S. 447–455

[62] GLOWATZKI, H. ; DUHM, S. ; BRAUN, K.-F. ; RABE, J. P. ; KOCH, N.: Molecular chains and carpets of sexithiophenes on Au(111). In: *Phys. Rev. B* 76 (2007), S. 125425

[63] MAKINEN, A. J. ; LONG, J. P. ; WATKINS, N. J. ; KAFAFI, Z. H.: Sexithiophene Adlayer Growth on Vicinal Gold Surfaces. In: *J. Phys. Chem. B* 109 (2005), Nr. 12, S. 5790–5795

[64] KOCH, N. ; VOLLMER, A. ; DUHM, S. ; SAKAMOTO, Y. ; SUZUKI, T.: The Effect of Fluorination on Pentacene/Gold Interface Energetics and Charge Reorganization Energy. In: *Adv. Mater.* 19 (2007), S. 112–116

[65] WITTE, G. ; WÖLL, Ch.: Molecular beam deposition and characterization of thin organic films on metals for applications in organic electronics. In: *Phys Status Solidi A* 205 (2008), Nr. 3, S. 497–510

[66] EREMTCHENKO, M. ; TEMIROV, R. ; BAUER, D. ; SCHAEFER, J. A. ; TAUTZ, F. S.: Formation of molecular order on a disordered interface layer: Pentacene/Ag(111). In: *Phys. Rev. B* 72 (2005), Nr. 11, S. 115430

[67] WAGNER, T. ; BANNANI, A. ; BOBISCH, C. ; KARACUBAN, H. ; MÖLLER, R.: The initial growth of PTCDA on Cu(111) studied by STM. In: *J Phys : Condens Matter* 19 (2007), Nr. 5, S. 056009. – ISSN 0953–8984

[68] TOERKER, M. ; FRITZ, T. ; PROEHL, H. ; SELLAM, F. ; LEO, K.: Tunneling spectroscopy study of 3,4,9,10-perylenetetracarboxylic dianhydride on Au(1 0 0). In: *Surf. Sci.* 491 (2001), Nr. 1-2, S. 255–264. – ISSN 0039–6028

Bibliography

[69] TAUTZ, F.S.: Structure and bonding of large aromatic molecules on noble metal surfaces: The example of PTCDA. In: *Prog Surf. Sci.* 82 (2007), Nr. 9-12, S. 479–520. – ISSN 0079–6816

[70] BURKE, S. A. ; TOPPLE, J. M. ; GRUTTER, P.: Molecular dewetting on insulators. In: *J Phys : Condens Matter* 21 (2009), Nr. 42, S. 423101. – ISSN 0953–8984

[71] GAO, J. ; XU, J. B. ; ZHU, M. ; KE, N. ; MA, Dongge: Thickness dependence of mobility in CuPc thin film on amorphous SiO2 substrate. In: *J Phys D: Appl Phys* 40 (2007), Nr. 18, S. 5666–5669. – ISSN 0022–3727

[72] LOI, M. A. ; DA COMO, E. ; DINELLI, F. ; MURGIA, M. ; ZAMBONI, R. ; BISCARINI, F. ; MUCCINI, M.: Supramolecular organization in ultra-thin α-sexithiophene on silicon dioxide. In: *Nature Mater.* 4 (2005), S. 81

[73] DINELLI, F. ; MOULIN, J.-F. ; LOI, M. A. ; DACOMO, E. ; MASSI, M. ; MURGIA, M. ; MUCCINI, M. ; BISCARINI, F. ; WIE, J. ; KINGSHOTT, P.: Effects of Surface Chemical Composition on the Early Growth Stages of α-Sexithienyl Films on Silicon Oxide Substrates. In: *J. Phys. Chem. B* 110 (2006), S. 258–263

[74] RUIZ, R. ; CHOUDHARY, D. ; NICKEL, B. ; TOCCOLI, T. ; CHANG, K.-C. ; MAYER, A.C. ; CLANCY, P. ; BLAKELY, J.M. ; HEADRICK, R.L. ; IANNOTTA, S. ; MALLIARAS, G.G.: Pentacene Thin Film Growth. In: *Chem. Mater.* 16 (2004), Nr. 23, S. 4497–4508

[75] GUNDLACH, D. J. ; NICHOLS, J. A. ; ZHOU, L. ; JACKSON, T. N.: Thin-film transistors based on well-ordered thermally evaporated naphthacene films. In: *Appl. Phys. Lett.* 80 (2002), Nr. 16, S. 2925–2927

[76] KÄFER, D. ; RUPPEL, L. ; WITTE, G.: Growth of pentacene on clean and modified gold surfaces. In: *Phys. Rev. B* 75 (2007), Nr. 8, S. 085309–14

[77] IHM, K. ; KIM, B. ; KANG, T.-H. ; KIM, K.-J. ; JOO, M. H. ; KIM, T. H. ; YOON, S. S. ; CHUNG, S.: Molecular orientation dependence of hole-injection barrier in pentacene thin film on the Au surface in organic thin film transistor. In: *Appl. Phys. Lett.* 89 (2006), Nr. 3, S. 033504–3

[78] HERTZ, H.: über einen Einfluss des ultravioletten Lichtes auf die electrische Entladung. In: *Ann. Phys.* 31 (1887), S. 983

[79] EINSTEIN, A.: über einen die Erzeugung und Verwandlung des Lichtes betreffenden heuristischen Gesichtspunkt. In: *Ann. Phys.* 17 (1905), Nr. 132, S. 20

Bibliography

[80] SEAH, M. P. ; DENCH, W. A.: Quantitative electron spectroscopy of surfaces: A standard data base for electron inelastic mean free paths in solids. In: *Surf. Interface Anal.* 1 (1979), Nr. 1, S. 2–11

[81] KOCH, N. ; POP, D. ; WEBER, R. L. ; BÖWERING, N. ; WINTER, B. ; WICK, M. ; LEISING, G. ; HERTEL, I. V. ; BRAUN, W.: Radiation induced degradation and surface charging of organic thin films in ultraviolet photoemission spectroscopy. In: *Thin Solid Films* 391 (2001), Nr. 1, S. 81–87. – ISSN 0040–6090

[82] KOCH, E.-E.: Photoemission from Organic Molecular Solids and Organometallic Compounds. In: *Phys Scr* T17 (1987), S. 120–136. – ISSN 1402–4896

[83] KOCH, N. ; DÜRR, A. C. ; GHIJSEN, J. ; JOHNSON, R. L. ; PIREAUX, J.-J. ; SCHWARTZ, J. ; SCHREIBER, F. ; DOSCH, H. ; KAHN, A.: Optically induced electron transfer from conjugated organic molecules to charged metal clusters. In: *Thin Solid Films* 441 (2003), Nr. 1-2, S. 145–149. – ISSN 0040–6090

[84] CAHEN, D. ; KAHN, A.: Electron Energetics at Surfaces and Interfaces: Concepts and Experiments. In: *Adv. Mater.* 15 (2003), S. 271–277

[85] GELIUS, U. ; HEDEN, P. F. ; HEDMAN, J. ; LINDBERG, B. J. ; MANNE, R. ; NORDBERG, R. ; NORDLING, C. ; SIEGBAHN, K.: Molecular Spectroscopy by Means of ESCA III. Carbon compounds. In: *Phys Scr* 2 (1970), Nr. 1-2, S. 70–80. – ISSN 0031–8949

[86] ALOV, N. V.: Fifty years of x-ray photoelectron spectroscopy. In: *J. Anal. Chem.* 60 (2005), Nr. 3, S. 297–300

[87] BRAUN, J.: The theory of angle-resolved ultraviolet photoemission and its applications to ordered materials. In: *Reports on Progress in Physics* 59 (1996), Nr. 10, S. 1267–1338. – ISSN 0034–4885

[88] BORSTEL, G.: Theoretical aspects of photoemission. In: *Applied Physics A: Materials Science & Processing* 38 (1985), Nr. 3, S. 193–204

[89] SPICER, W. E.: Photoemissive, Photoconductive, and Optical Absorption Studies of Alkali-Antimony Compounds. In: *Phys. Rev.* 112 (1958), Nr. 1, S. 114

[90] CARDONA, M. ; LEY, L.: *Photoemission in solids: General principles.* Springer-Verlag Berlin, Heidelberg, New York, 1978

[91] ERTL, G. ; J.KÜPPERS: *Low energy electrons and surface chemistry.* VCH Weinheim, 1985

Bibliography

[92] WILLOCK, D.: *Molecular Symmetry*. Wiley, 2009

[93] THORNE, A.P. ; LITZÉN, U. ; JOHANSSON, S.: *Spectrophysics: Principles and Applications*. Springer Verlag Berlin Heidelberg, 1999

[94] LÜTH, H.: *Solid surfaces, interfaces and thin films*. Springer Verlag, 2001

[95] YAMANE, H. ; NAGAMATSU, S. ; FUKAGAWA, H. ; KERA, S. ; FRIEDLEIN, R. ; OKUDAIRA, K. K. ; UENO, N.: Hole-vibration coupling of the highest occupied state in pentacene thin films. In: *Phys. Rev. B* 72 (2005), S. 153412

[96] HIMPSEL, F. J.: Angle-resolved measurements of the photoemission of electrons in the study of solids. In: *Adv. Phys.* 32 (1983), S. 1–51

[97] KOCH, N. ; HEIMEL, G. ; WU, J. ; ZOJER, E. ; JOHNSON, R. L. ; BRÉDAS, J.-L. ; MÜLLEN, K. ; RABE, J. P.: Influence of molecular conformation on organic/metal interface energetics. In: *Chem. Phys. Lett.* 413 (2005), Nr. 4-6, S. 390–395

[98] BRÖKER, B. ; BLUM, R.-P. ; BEVERINA, L. ; HOFMANN, O. T. ; SASSI, M. ; RUFFO, R. ; PAGANI, G. A. ; HEIMEL, G. ; VOLLMER, A. ; FRISCH, J. ; RABE, J. P. ; ZOJER, E. ; KOCH, N.: A High Molecular Weight Donor for Electron Injection Interlayers on Metal Electrodes. In: *Chem. Phys. Chem.* 10 (2009), Nr. 17, S. 2947 – 2954

[99] KUZMANY, H.: *Solid-State Spectroscopy: an Introduction*. Springer Verlag Berlin Heidelberg, 2009

[100] NEEDHAM, Jr. ; DRISCOLL, T.J. ; RAO, N.G.: Correlation of fractional-monolayer oxygen determinations obtained by proton-excited x-ray and Auger electron spectroscopy analysis of Fe surfaces. In: *Appl. Phys. Lett.* 21 (1972), Nr. 10, S. 502–505

[101] BRIGGS, D. ; SEAH, M. P.: *Practical surface analysis by Auger and X-ray photoelectron spectroscopy*. John Wiley & Sons, 1983

[102] HAKEN, H. ; WOLF, H.C.: *Molekülphysik und Quantenchemie*. Springer-Verlag Berlin Heidelberg, 2006

[103] HAKEN, H. ; WOLF, H.C.: *Atom-und Quantenphysik*. Springer Berlin Heidelberg, 1980

[104] MERLIN, Jean-Claude ; CORNARD, Jean-Paul: A Pictorial Representation of Normal Modes of Vibration Using Vibrational Symmetry Coordinates. In: *J Chem Educ* 83 (2006), Nr. 9, S. 1393

[105] YATES, J.T. ; MADEY, T.E.: *Vibrational spectroscopy of molecules on surfaces*. Plenum Press New York, 1987

Bibliography

[106] HOFFMANN, F. M.: Infrared reflection-absorption spectroscopy of adsorbed molecules. In: *Surf. Sci. Rep.* 3 (1983), S. 107

[107] FRANCIS, S. A. ; ELLISON, A. H.: Infrared spectra of monolayers on metal mirrors. In: *Journal of the Optical Society of America* 49 (1959), Nr. 2, S. 131–137

[108] GREENLER, R. G.: Reflection Method for Obtaining the Infrared Spectrum of a Thin Layer on a Metal Surface. In: *J Chem. Phys.* 50 (1969), Nr. 5, S. 1963–1968

[109] MCINTYRE, J.D.E. ; ASPNES, D.E.: Differential reflection spectroscopy of very thin surface films. In: *Surf. Sci.* 24 (1971), S. 417–434

[110] HOLLINS, P. ; PRITCHARD, J.: Infrared studies of chemisorbed layers on single crystals. In: *Prog Surf. Sci.* 19 (1985), Nr. 4, S. 275–349

[111] GRIFFITHS, P. R. ; HASETH, J. A. D.: *Fourier transform infrared spectrometry*. Wiley-Interscience, 2007

[112] HOLLINS, P.: Surface infrared spectroscopy. In: *Vacuum* 45 (1994), Nr. 6-7, S. 705–714. – ISSN 0042–207X

[113] IBACH, H. ; MILLS, DL: *Electron energy loss spectroscopy and surface vibrations*. Academic Press New York, 1982

[114] HAQ, S. ; KING, DA: Configurational Transitions of Benzene and Pyridine Adsorbed on Pt {111} and Cu {110} Surfaces: An Infrared Study. In: *J. Phys. Chem* 100 (1996), Nr. 42, S. 16957–16965

[115] EFRIMA, S. ; METIU, H.: Vibrational frequencies of a chemisorbed molecule: The role of the electrodynamic interactions. In: *Surf. Sci.* 92 (1980), Nr. 2-3, S. 433–452. – ISSN 0039–6028

[116] EFRIMA, S. ; METIU, H.: The change of the vibrational frequencies of a chemisorbed diatomic caused by the electrostatic interactions with the metal surface. In: *Surf. Sci.* 108 (1981), Nr. 2, S. 329–339. – ISSN 0039–6028

[117] BLYHOLDER, G.: Molecular orbital view of chemisorbed carbon monoxide. In: *J. Phys. Chem.* 68 (1964), Nr. 10, S. 2772–2777

[118] KHATKALE, M. S. ; DEVLIN, J. P.: The vibrational and electronic spectra of the mono-, di-, and trianon salts of TCNQ. In: *J Chem. Phys.* 70 (1979), Nr. 4, S. 1851–1859

[119] BRADSHAW, A. M. ; SCHEFFLER, M.: Lateral interactions in adsorbed layers. In: *Journal of Vacuum Science and Technology* 16 (1979), Nr. 2, S. 447–454

Bibliography

[120] HAMMETT, L. P.: The Effect of Structure upon the Reactions of Organic Compounds. Benzene Derivatives. In: *J. Am. Chem. Soc.* 59 (1937), Nr. 1, S. 96–103

[121] HANSCH, C. ; LEO, A. ; TAFT, R. W.: A survey of Hammett substituent constants and resonance and field parameters. In: *Chem. Rev.* 91 (1991), Nr. 2, S. 165–195

[122] KATO, T. ; MORI, T. ; MIZUTANI, T.: Effect of fabrication conditions on photoluminescence and absorption of hole transport materials. In: *Thin Solid Films* 393 (2001), Nr. 1-2, S. 109–113. – ISSN 0040-6090

[123] SUGIYAMA, T. ; FURUKAWA, Y. ; FUJIMURA, H.: Crystalline/amorphous Raman markers of hole-transport material NPD in organic light-emitting diodes. In: *Chem. Phys. Lett.* 405 (2005), Nr. 4-6, S. 330–333. – ISSN 0009-2614

[124] BRINKMANN, M. ; GADRET, G. ; MUCCINI, M. ; TALIANI, C. ; MASCIOCCHI, N. ; SIRONI, A.: Correlation between Molecular Packing and Optical Properties in Different Crystalline Polymorphs and Amorphous Thin Films of mer-Tris(8-hydroxyquinoline)aluminum(III). In: *J. Am. Chem. Soc.* 122 (2000), Nr. 21, S. 5147–5157

[125] ALTMAN, E. I. ; COLTON, R. J.: The interaction of C60 with noble metal surfaces. In: *Surf. Sci.* 295 (1993), Nr. 1-2, S. 13–33. – ISSN 0039-6028

[126] TSUEI, K. D. ; YUH, J. Y. ; TZENG, C. T. ; CHU, R. Y. ; CHUNG, S. C. ; TSANG, K. L.: Photoemission and photoabsorption study of C_ $$60$$ adsorption on Cu (111) surfaces. In: *Phys. Rev. B* 56 (1997), Nr. 23, S. 15412–15420

[127] JOHNSON, R. L. ; REICHARDT, J.: FLIPPER II – a new photoemission system in HASYLAB. In: *Nucl. Instrum. Methods Phys. Res.* 208 (1983), S. 791–796

[128] EICHLER, J.: *Laser: Bauformen, Strahlführung, Anwendungen*. Springer, 2003

[129] BARNES, R. B. ; BONNER, Lyman G.: The Christiansen Filter Effect in the Infrared. In: *Phys. Rev.* 49 (1936), Nr. 10, S. 732–740

[130] PRIMAS, H. ; GÜNTHARD, H. H.: Theorie der Form von Absorptionsbanden suspendierter Substanzen und deren Anwendung auf die Nujolmethode in der Infrarotspektroskopie. In: *Helv Chim Acta* 37 (1954), Nr. 1, S. 360–374

[131] HOHENBERG, P. ; KOHN, W.: Inhomogeneous Electron Gas. In: *Phys. Rev.* 136 (1964), Nr. 3B, S. B864

[132] KOHN, W. ; SHAM, L. J.: Self-Consistent Equations Including Exchange and Correlation Effects. In: *Phys. Rev.* 140 (1965), Nr. 4A, S. A1133

Bibliography

[133] KOHN, W. ; BECKE, A. D. ; PARR, R. G.: Density Functional Theory of Electronic Structure. In: *J. Phys. Chem.* 100 (1996), Nr. 31, S. 12974–12980

[134] KRESSE, G. ; HAFNER, J.: Ab initio molecular dynamics for open-shell transition metals. In: *Phys. Rev. B* 48 (1993), Nr. 17, S. 13115

[135] KRESSE, G. ; HAFNER, J.: Ab initio molecular-dynamics simulation of the liquid-metal–amorphous-semiconductor transition in germanium. In: *Phys. Rev. B* 49 (1994), Nr. 20, S. 14251

[136] KRESSE, G. ; FURTHMÜLLER, J.: Efficiency of ab-initio total energy calculations for metals and semiconductors using a plane-wave basis set. In: *Comput. Mat. Sci.* 6 (1996), Nr. 1, S. 15–50

[137] KRESSE, G. ; FURTHMÜLLER, J.: Efficient iterative schemes for ab initio total-energy calculations using a plane-wave basis set. In: *Phys. Rev. B* 54 (1996), Nr. 16, S. 11169

[138] KRESSE, G. ; JOUBERT, D.: From ultrasoft pseudopotentials to the projector augmented-wave method. In: *Phys. Rev. B* 59 (1999), Nr. 3, S. 1758

[139] BARTH, J. V. ; BRUNE, H. ; ERTL, G. ; BEHM, R. J.: Scanning tunneling microscopy observations on the reconstructed Au(111) surface: Atomic structure, long-range superstructure, rotational domains, and surface defects. In: *Phys. Rev. B* 42 (1990), Nr. 15, S. 9307–9318

[140] PERDEW, J. P. ; CHEVARY, J. A. ; VOSKO, S. H. ; JACKSON, Koblar A. ; PEDERSON, M. R. ; SINGH, D. J. ; FIOLHAIS, C.: Atoms, molecules, solids, and surfaces: Applications of the generalized gradient approximation for exchange and correlation. In: *Phys. Rev. B* 46 (1992), Nr. 11, S. 6671

[141] BLÖCHL, P. E.: Projector augmented-wave method. In: *Phys. Rev. B* 50 (1994), Nr. 24, S. 17953

[142] MONKHORST, H. J. ; PACK, J. D.: Special points for Brillouin-zone integrations. In: *Phys. Rev. B* 13 (1976), Nr. 12, S. 5188

[143] METHFESSEL, M. ; PAXTON, A. T.: High-precision sampling for Brillouin-zone integration in metals. In: *Phys. Rev. B* 40 (1989), Nr. 6, S. 3616

[144] HEIMEL, G. ; ROMANER, L. ; BRÉDAS, J.-L. ; ZOJER, E.: Organic/metal interfaces in self-assembled monolayers of conjugated thiols: A first-principles benchmark study. In: *Surf. Sci.* 600 (2006), Nr. 19, S. 4548–4562

Bibliography

[145] FRISCH, M. J. ; TRUCKS, G. W. ; SCHLEGEL, H. B. ; SCUSERIA, G. E. ; ROBB, M. A. ; CHEESEMAN, J. R. ; MONTGOMERY, J. A. Jr. ; VREVEN, T. ; KUDIN, K. N. ; BURANT, J. C. ; MILLAM, J. M. ; IYENGAR, S. S. ; TOMASI, J. ; BARONE, V. ; MENNUCCI, B. ; COSSI, M. ; SCALMANI, G. ; REGA, N. ; PETERSSON, G. A. ; NAKATSUJI, H. ; HADA, M. ; EHARA, M. ; TOYOTA, K. ; FUKUDA, R. ; HASEGAWA, J. ; ISHIDA, M. ; NAKAJIMA, T. ; HONDA, Y. ; KITAO, O. ; NAKAI, H. ; KLENE, M. ; LI, X. ; KNOX, J. E. ; HRATCHIAN, H. P. ; CROSS, J. B. ; BAKKEN, V. ; ADAMO, C. ; JARAMILLO, J. ; GOMPERTS, R. ; STRATMANN, R. E. ; YAZYEV, O. ; AUSTIN, A. J. ; CAMMI, R. ; POMELLI, C. ; OCHTERSKI, J. W. ; AYALA, P. Y. ; MOROKUMA, K. ; VOTH, G. A. ; SALVADOR, P. ; DANNENBERG, J. J. ; ZAKRZEWSKI, V. G. ; DAPPRICH, S. ; DANIELS, A. D. ; STRAIN, M. C. ; FARKAS, O. ; MALICK, D. K. ; RABUCK, A. D. ; RAGHAVACHARI, K. ; FORESMAN, J. B. ; ORTIZ, J. V. ; CUI, Q. ; BABOUL, A. G. ; CLIFFORD, S. ; CIOSLOWSKI, J. ; STEFANOV, B. B. ; LIU, G. ; LIASHENKO, A. ; PISKORZ, P. ; KOMAROMI, I. ; MARTIN, R. L. ; FOX, D. J. ; KEITH, T. ; AL-LAHAM, M. A. ; PENG, C. Y. ; NANAYAKKARA, A. ; CHALLACOMBE, M. ; GILL, P. M. W. ; JOHNSON, B. ; CHEN, W. ; WONG, M. W. ; GONZALEZ, C. ; POPLE, J. A.: *Gaussian 03, Revision C.02*. 2004

[146] BECKE, A. D.: A new mixing of Hartree–Fock and local density-functional theories. In: *J Chem. Phys.* 98 (1993), Nr. 2, S. 1372–1377

[147] LEE, C. ; YANG, W. ; PARR, R. G.: Development of the Colle-Salvetti correlation-energy formula into a functional of the electron density. In: *Phys. Rev. B* 37 (1988), Nr. 2, S. 785

[148] HARIHARAN, P. C. ; POPLE, J. A.: The influence of polarization functions on molecular orbital hydrogenation energies. In: *Theoretical Chemistry Accounts: Theory, Computation, and Modeling (Theoretica Chimica Acta)* 28 (1973), Nr. 3, S. 213–222

[149] GAO, Z. Q. ; MI, B. X. ; XU, G. Z. ; WAN, Y. Q. ; GONG, M. L. ; CHEAH, K. W. ; CHEN, C. H.: An organic p-type dopant with high thermal stability for an organic semiconductor. In: *Chem. Commun.* 1 (2008), Nr. 1, S. 117–119

[150] MONK, P.M.S.: *The viologens: physicochemical properties, synthesis and applications of the salts of 4, 4'-bipyridine*. John Wiley & Sons, 1999

[151] BRÖKER, B. ; BLUM, R.-P. ; FRISCH, J. ; VOLLMER, A. ; HOFMANN, O. T. ; RIEGER, R. ; MÜLLEN, K. ; RABE, J. P. ; ZOJER, E. ; KOCH, N.: Gold work function reduction by 2.2 eV with an air-stable molecular donor layer. In: *Appl. Phys. Lett.* 93 (2008), Nr. 24, S. 243303

Bibliography

[152] PALMGREN, P. ; YU, S. ; HENNIES, F. ; NILSON, K. ; AKERMARK, B. ; GÖTHELID, M.: Changing adsorption mode of FePc on TiO2(110) by surface modification with bipyridine. In: *J Chem. Phys.* 129 (2008), Nr. 7, S. 074707. – ISSN 00219606

[153] SCHREIBER, F.: Structure and growth of self-assembling monolayers. In: *Prog Surf. Sci.* 65 (2000), Nr. 5-8, S. 151–257. – ISSN 0079–6816

[154] HOFMANN, O. T. ; RANGGER, G. M. ; ZOJER, E.: Reducing the Metal Work Function beyond Pauli Pushback: A Computational Investigation of Tetrathiafulvalene and Viologen on Coinage Metal Surfaces. In: *J. Phys. Chem. C* 112 (2008), Nr. 51, S. 20357–20365

[155] BRAUN, S. ; SALANECK, W. R.: Fermi level pinning at interfaces with tetrafluorotetracyanoquinodimethane (F4-TCNQ): The role of integer charge transfer states. In: *Chem. Phys. Lett.* 438 (2007), Nr. 4-6, S. 259–262

[156] POIZAT, O. ; SOURISSEAU, C. ; CORSET, J.: Vibrational and electronic study of the methyl viologen radical cation MV+. in the solid state. In: *J Mol Struct* 143 (1986), S. 203–206. – ISSN 0022–2860

[157] ITO, M. ; SASAKI, H. ; TAKAHASHI, M.: Infrared spectra and dimer structure of reduced viologen compounds. In: *J. Phys. Chem.* 91 (1987), Nr. 15, S. 3932–3934

[158] CHRISTENSEN, P.A. ; HAMNETT, A.: An in-situ FTIR study into the nature of completely reduced MV2+. In: *J Electroanal Chem* 263 (1989), Nr. 1, S. 49–68. – ISSN 0022–0728

[159] ARIHARA, K. ; KITAMURA, F.: Adsorption states of heptyl viologen on an Au(111) electrode surface studied by infrared reflection absorption spectroscopy. In: *J Electroanal Chem* 550-551 (2003), S. 149–159

[160] HUANG, W. X. ; WHITE, J. M.: Propene adsorption on Ag(1 1 1): a TPD and RAIRS study. In: *Surf. Sci.* 513 (2002), Nr. 2, S. 399–404. – ISSN 0039–6028

[161] HUANG, W. X. ; WHITE, J. M.: Growth and Orientation of Naphthalene Films on Ag(111). In: *J. Phys. Chem. B* 108 (2004), Nr. 16, S. 5060–5065

[162] MARDER, S.R. ; KIPPELEN, B. ; ALEX, K.Y.J. ; PEYGHAMBARIAN, N.: Design and synthesis of chromophores and polymers for electro-optic and photorefractive applications. In: *Nature* 388 (1997), Nr. 6645, S. 845–851

[163] HILL, I. G. ; MÄKINEN, A. J. ; KAFAFI, Z. H.: Initial stages of metal/organic semiconductor interface formation. In: *J. Appl. Phys.* 88 (2000), S. 889–895

Bibliography

[164] LOF, R. W. ; VEENENDAAL, M. A. ; KOOPMANS, B. ; JONKMAN, H. T. ; SAWATZKY, G. A.: Band gap, excitons, and Coulomb interaction in solid C60. In: *Phys. Rev. Lett.* 68 (1992), Nr. 26, S. 3924

[165] BENNING, P. J. ; POIRIER, D. M. ; OHNO, T. R. ; CHEN, Y. ; JOST, M. B. ; STEPNIAK, F. ; KROLL, G. H. ; WEAVER, J. H. ; FURE, J. ; SMALLEY, R. E.: C60 and C70 fullerenes and potassium fullerides. In: *Phys. Rev. B* 45 (1992), Nr. 12, S. 6899

[166] VEENSTRA, S.C. ; HEERES, A. ; HADZIIOANNOU, G. ; SAWATZKY, G.A. ; JONKMAN, H.T.: On interface dipole layers between C60 and Ag or Au. In: *Appl. Phys. A* 75 (2002), Nr. 6, S. 661–666

[167] GLOWATZKI, H. ; BRÖKER, B. ; BLUM, R.-P. ; HOFMANN, O. T. ; VOLLMER, A. ; RIEGER, R. ; MÜLLEN, K. ; ZOJER, E. ; RABE, J. P. ; KOCH, N.: "Soft" Metallic Contact to Isolated C60 Molecules. In: *Nano. Lett.* 8 (2008), Nr. 11, S. 3825–3829

[168] TZENG, C.-T. ; LO, W.-S. ; YUH, J.-Y. ; CHU, R.-Y. ; TSUEI, K.-D.: Photoemission, near-edge x-ray-absorption spectroscopy, and low-energy electron-diffraction study of C60 on Au(111) surfaces. In: *Phys. Rev. B* 61 (2000), Nr. 3, S. 2263

[169] RAJAGOPAL, A. ; KAHN, A.: Photoemission spectroscopy investigation of magnesium–Alq[sub 3] interfaces. In: *J. Appl. Phys.* 84 (1998), Nr. 1, S. 355–358

[170] NIELSEN, S. B. ; NIELSEN, M. B. ; JENSEN, H. J. A.: The tetrathiafulvalene dication in the gas phase: its formation and stability. In: *Phys Chem Chem. Phys.* 5 (2003), Nr. 7, S. 1376–1380

[171] GROBMAN, W. D. ; POLLAK, R. A. ; EASTMAN, D. E. ; MAAS, E. T. ; SCOTT, B. A.: Valence Electronic Structure and Charge Transfer in Tetrathiofulvalinium Tetracyanoquinodimethane (TTF-TCNQ) from Photoemission Spectroscopy. In: *Phys. Rev. Lett.* 32 (1974), Nr. 10, S. 534

[172] FERNANDEZ-TORRENTE, I. ; MONTURET, S. ; FRANKE, K.J. ; FRAXEDAS, J. ; LORENTE, N. ; PASCUAL, J.I.: Long-Range Repulsive Interaction between Molecules on a Metal Surface Induced by Charge Transfer. In: *Phys. Rev. Lett.* 99 (2007), Nr. 17, S. 176103

[173] KÄFER, D. ; RUPPEL, L. ; WITTE, G. ; WÖLL, C.: Role of Molecular Conformations in Rubrene Thin Film Growth. In: *Phys. Rev. Lett.* 95 (2005), Nr. 16, S. 166602

[174] HÜCKSTÄDT, C. ; SCHMIDT, S. ; HÜFNER, S. ; FORSTER, F. ; REINERT, F. ; SPRINGBORG, M.: Work function studies of rare-gas/noble metal adsorption systems using a Kelvin probe. In: *Phys. Rev. B* 73 (2006), Nr. 7, S. 75409

Bibliography

[175] NARIOKA, S. ; ISHII, H. ; YOSHIMURA, D. ; SEI, M. ; OUCHI, Y. ; SEKI, K. ; HASEGAWA, S. ; MIYAZAKI, T. ; HARIMA, Y. ; YAMASHITA, K.: The electronic structure and energy level alignment of porphyrin/metal interfaces studied by ultraviolet photoelectron spectroscopy. In: *Appl. Phys. Lett.* 67 (1995), Nr. 13, S. 1899–1901

[176] JAECKEL, B. ; SAMBUR, J. B. ; PARKINSON, B. A.: The influence of metal work function on the barrier heights of metal/pentacene junctions. In: *J. Appl. Phys.* 103 (2008), Nr. 6, S. 063719-7

[177] DUHM, S. ; GLOWATZKI, H. ; CIMPEANU, V. ; KLANKERMAYER, J. ; RABE, J. P. ; JOHNSON, R. L. ; KOCH, N.: Weak Charge Transfer between an Acceptor Molecule and Metal Surfaces Enabling Organic/Metal Energy Level Tuning. In: *J. Phys. Chem. B* 110 (2006), Nr. 42, S. 21069–21072

[178] WERNER, W. S. M.: Electron transport in solids for quantitative surface analysis. In: *Surf. Interface Anal.* 31 (2001), Nr. 3, S. 141–176

[179] SITAR, Z. ; SMITH, L. L. ; DAVIS, R. F.: Interface chemistry and surface morphology in the initial stages of growth of GaN and AlN on [alpha]-SiC and sapphire. In: *J Cryst Growth* 141 (1994), Nr. 1-2, S. 11–21. – ISSN 0022-0248

[180] RANGGER, G. M. ; HOFMANN, O. T. ; ROMANER, L. ; HEIMEL, G. ; BRÖKER, B. ; BLUM, R.-P. ; JOHNSON, R. L. ; KOCH, N. ; ZOJER, E.: F4TCNQ on Cu, Ag, and Au as prototypical example for a strong organic acceptor on coinage metals. In: *Phys. Rev. B* 79 (2009), Nr. 16, S. 165306-12

[181] JÄCKEL, F. ; PERERA, U. G. E. ; IANCU, V. ; BRAUN, K.-F. ; KOCH, N. ; RABE, J. P. ; HLA, S.-W.: Investigating Molecular Charge Transfer Complexes with a Low Temperature Scanning Tunneling Microscope. In: *Phys. Rev. Lett.* 100 (2008)

[182] TAUTZ, F. S. ; EREMTCHENKO, M. ; SCHAEFER, J. A. ; SOKOLOWSKI, M. ; SHKLOVER, V. ; UMBACH, E.: Strong electron-phonon coupling at a metal/organic interface: PTCDA/Ag(111). In: *Phys. Rev. B* 65 (2002), Nr. 12, S. 125405

[183] TEMIROV, R. ; SOUBATCH, S. ; LASSISE, A. ; TAUTZ, F. S.: Bonding and vibrational dynamics of a large pi-conjugated molecule on a metal surface. In: *J Phys : Condens Matter* 20 (2008), Nr. 22, S. 224010. – ISSN 0953-8984

[184] KAMITSOS, E. I. ; RISEN, J.: Raman studies in CuTCNQ: Resonance Raman spectral observations and calculations for TCNQ ion radicals. In: *J Chem. Phys.* 79 (1983), Nr. 12, S. 5808–5819

[185] ERLEY, W. ; IBACH, H.: Vibrational spectra of tetracyanoquinodimethane (TCNQ) adsorbed on the Cu(111) surface. In: *Surf. Sci.* 178 (1986), Nr. 1-3, S. 565–577. – ISSN 0039-6028

[186] MENEGHETTI, M. ; PECILE, C.: Charge–transfer organic crystals: Molecular vibrations and spectroscopic effects of electron–molecular vibration coupling of the strong electron acceptor TCNQF4. In: *J Chem. Phys.* 84 (1986), Nr. 8, S. 4149–4162

[187] LU, J. ; LOH, K. P.: High resolution electron energy loss spectroscopy study of Zinc phthalocyanine and tetrafluoro tetracyanoquinodimethane on Au (1 1 1). In: *Chem. Phys. Lett.* 468 (2009), Nr. 1-3, S. 28–31. – ISSN 0009-2614

[188] KRAMER, M. ; HOFFMANN, V.: Infrared spectroscopic characterization of orientation and order of thin oligothiophene films. In: *Optical Materials* 9 (1998), Nr. 1-4, S. 65–69. – ISSN 0925-3467

[189] BRÖKER, B. ; HOFMANN, O. T. ; RANGGER, G. M. ; FRANK, P. ; BLUM, R.-P. ; RIEGER, R. ; VENEMA, L. ; VOLLMER, A. ; MÜLLEN, K. ; RABE, J. P. ; WINKLER, A. ; RUDOLF, P. ; ZOJER, E. ; KOCH, N.: Density-Dependent Reorientation and Rehybridization of Chemisorbed Conjugated Molecules for Controlling Interface Electronic Structure. In: *Phys. Rev. Lett.* 104 (2010), Juni, Nr. 24, S. 246805

[190] VÁZQUEZ, H. ; OSZWALDOWSKI, R. ; POU, P. ; ORTEGA, J. ; PÉREZ, R. ; FLORES, F. ; KAHN, A.: Dipole formation at metal/PTCDA interfaces: Role of the Charge Neutrality Level. In: *Europhys. Lett.* 65 (2004), S. 802–808

[191] BAGUS, P. S. ; HERMANN, K. ; WÖLL, C.: The interaction of C_6H_6 and C_6H_{12} with noble metal surfaces: Electronic level alignment and the origin of the interface dipole. In: *J Chem. Phys.* 123 (2005), S. 184109

[192] HAUSCHILD, A. ; KARKI, K. ; COWIE, B. C. C. ; ROHLFING, M. ; TAUTZ, F. S. ; SOKOLOWSKI, M.: Molecular Distortions and Chemical Bonding of a Large pi-Conjugated Molecule on a Metal Surface. In: *Phys. Rev. Lett.* 94 (2005), S. 036106

[193] STADLER, C. ; HANSEN, S. ; KRÖGER, I. ; KUMPF, C. ; UMBACH, E.: Tuning intermolecular interaction in long-range-ordered submonolayer organic films. In: *Nature Phys.* 5 (2009), Nr. 2, S. 153–158. – ISSN 1745-2473

[194] HUANG, Q. ; EVMENENKO, G. A. ; DUTTA, P. ; LEE, P. ; ARMSTRONG, N. R. ; MARKS, T. J.: Covalently Bound Hole-Injecting Nanostructures. Systematics of Molecular Architecture, Thickness, Saturation, and Electron-Blocking Characteristics on Organic

Light-Emitting Diode Luminance, Turn-on Voltage, and Quantum Efficiency. In: *J. Am. Chem. Soc.* 127 (2005), Nr. 29, S. 10227–10242

[195] FORKER, R. ; GOLNIK, C. ; PIZZI, G. ; DIENEL, T. ; FRITZ, T.: Optical absorption spectra of ultrathin PTCDA films on gold single crystals: Charge transfer beyond the first monolayer. In: *Org. Electron.* 10 (2009), Nr. 8, S. 1448–1453. – ISSN 1566–1199

[196] BRIDGE, M.E ; MARBROW, R.A ; LAMBERT, R.M: Adsorption of cyanogen and hydrogen cyanide on Pt(110) and Ag(110). In: *Surf. Sci.* 57 (1976), Nr. 1, S. 415–419. – ISSN 0039–6028

[197] LANGRETH, D. C.: Energy Transfer at Surfaces: Asymmetric Line Shapes and the Electron-Hole-Pair Mechanism. In: *Phys. Rev. Lett.* 54 (1985), Nr. 2, S. 126

[198] SZALAY, P. S. ; GALÁN-MASCARÓS, J. R. ; CLÉRAC, R. ; DUNBAR, K. R.: HAT(CN)6: a new building block for molecule-based magnetic materials. In: *Synth. Met.* 122 (2001), Nr. 3, S. 535–542. – ISSN 0379–6779

[199] HEIMEL, G. ; ROMANER, L. ; BREDAS, J.-L. ; ZOJER, E.: Interface Energetics and Level Alignment at Covalent Metal-Molecule Junctions: pi-Conjugated Thiols on Gold. In: *Phys. Rev. Lett.* 96 (2006), Nr. 19, S. 196806–4

[200] WEAVER, J.H.: Electronic structures of C60, C70 and the fullerides: Photoemission and inverse photoemission studies. In: *J. Phys. Chem. Solids* 53 (1992), Nr. 11, S. 1433–1447. – ISSN 0022–3697

[201] AMSALEM, P.: *Private Communication.* 2010

[202] FRANK, P. ; KOCH, N. ; KOINI, M. ; RIEGER, R. ; MÜLLEN, K. ; RESEL, R. ; WINKLER, A.: Layer growth and desorption kinetics of a discoid molecular acceptor on Au(1 1 1). In: *Chem. Phys. Lett.* 473 (2009), S. 321–325. – ISSN 0009–2614

[203] HILL, I. G. ; KAHN, A.: Energy level alignment at interfaces of organic semiconductor heterostructures. In: *J. Appl. Phys.* 84 (1998), S. 5583–5586

[204] OSIKOWICZ, W. ; DE JONG, M. P. ; SALANECK, W. R.: Formation of the Interfacial Dipole at Organic-Organic Interfaces: C60/Polymer Interfaces. In: *Adv. Mater.* 19 (2007), Nr. 23, S. 4213–4217

[205] HILL, I. G. ; KAHN, A.: Organic semiconductor heterointerfaces containing bathocuproine. In: *J. Appl. Phys.* 86 (1999), Nr. 8, S. 4515–4519

[206] HENZLER, M. ; GÖPEL, W.: *Oberflächenphysik des Festkörpers.* B. G. Teubner Stuttgart, 1994

Bibliography

[207] HEINZE, J.: Cyclovoltammetrie die Spektroskopie des Elektrochemikers. In: *Angew. Chem.* 96 (1984), Nr. 11, S. 823–840

[208] BARD, A. J. ; FAULKNER, L. R.: *Electrochemical Methods. Fundamentals and Applications*, J. Wiley & Sons, New York, 1980

[209] BARRETTE, W. C. ; JOHNSON, H. W. ; SAWYER, D. T.: Voltammetric evaluation of the effective acidities (pKa') for Broensted acids in aprotic solvents. In: *Anal. Chem.* 56 (1984), Nr. 11, S. 1890–1898

[210] KRÖGER, J.: Electron-phonon coupling at metal surfaces. In: *Rep. Prog. Phys.* 69 (2006), Nr. 4, S. 899–969. – ISSN 0034–4885

[211] BAUER, E. ; POPPA, H.: Recent advances in epitaxy. In: *Thin Solid Films* 12 (1972), Nr. 1, S. 167–185. – ISSN 0040–6090

[212] BAUER, E. ; POPPA, H. ; TODD, G. ; BONCZEK, F.: Adsorption and condensation of Cu on W single-crystal surfaces. In: *J. Appl. Phys.* 45 (1974), Nr. 12, S. 5164–5175

Die VDM Verlagsservicegesellschaft sucht für wissenschaftliche Verlage abgeschlossene und herausragende

Dissertationen, Habilitationen, Diplomarbeiten, Master Theses, Magisterarbeiten usw.

für die kostenlose Publikation als Fachbuch.

Sie verfügen über eine Arbeit, die hohen inhaltlichen und formalen Ansprüchen genügt, und haben Interesse an einer honorarvergüteten Publikation?

Dann senden Sie bitte erste Informationen über sich und Ihre Arbeit per Email an *info@vdm-vsg.de*.

Sie erhalten kurzfristig unser Feedback!

VDM Verlagsservicegesellschaft mbH
Dudweiler Landstr. 99   Telefon +49 681 3720 174
D - 66123 Saarbrücken   Fax   +49 681 3720 1749
www.vdm-vsg.de

Die VDM Verlagsservicegesellschaft mbH vertritt

Printed by Books on Demand GmbH, Norderstedt / Germany